○ ● ○ ○

Mathematics
in Daily Living
REVISED

DECIMALS and PERCENT

Nerissa Bell Bryant
Staff Consultant
Adult Performance Level Project
Louisiana Tech University

Loy Hedgepeth
Director of Adult Education
Ouachita Parish Schools
Monroe, Louisiana

Steck-Vaughn Company ○ Austin, Texas

Introduction

This book contains adult-oriented instructional material designed to teach mathematics skills and life-coping skills to the mature learner.

The academic skill content of this book was determined by data compiled from a survey of adult education teachers. This survey revealed the topics for which teachers most urgently needed teaching materials.

In addition to the mathematics skill content, this book focuses on life-coping skills needed daily by adults. The curriculum emphasis focuses on these five general areas: health, government and law, consumer economics, community resources, and occupational knowledge. Information on various topics in these five areas is presented along with practice of mathematics skills. Thus, as learners progress through the mathematics units, they are afforded an opportunity to (1) develop academic skills related to *mathematics competence* and (2) develop life-coping skills related to *functional competence*.

During their development these materials were field-tested at the Northeast Louisiana Learning Center in Monroe, Louisiana.

How To Use This Book

Each unit contains individualized instruction for self-development. However, the format and use of the book should be understood in order to expedite progress and maximize proper use of the material.

Unit: A section providing instruction, examples, practices, reviews, and evaluations of a designated skill. Units are subdivided into lessons.

Skill: Each intermediate skill necessary to the mastery of the designated unit skill is presented in each lesson title.

Instruction: Explanation and/or definition of each skill is offered. Examples follow each segment of instruction to clarify the instruction and prepare the learner to undertake practice exercises.

Exercises: Exercises to be worked and checked by the learner. Exercise B provides the same type of practice as Exercise A.

Review: An exercise providing practice and review on skills taught in the unit.

Coping Skills: Most units end with an activity designed to develop greater competency in life-coping skills. Other activities related to coping skills are within the lessons.

Answers: Answers to all pretests, exercises, and reviews are provided at the back of the book.

NOTICE: Answer Key is bound in the back of the book.

ISBN 0-8114-1514-7

5 6 7 8 9 0 C 94 93 92 91 90

Contents

UNIT 1—PRETEST
READING AND WRITING DECIMALS

Write these decimals as fractions.

1. .08 = ...   2. .11 = ...

3. .025 = ...   4. .983 = ...

5. .2402 = ...   6. .7153 = ...

7. .00001 = ...   8. .56789 = ...

Write each of the following in decimal form and fractional form.

9. 71 ten-thousandths ....

10. 2,004 hundred-thousandths ....

11. 8 tenths ....

12. 5 thousandths ....

Find the decimal form and fraction form of each of the following. Blacken the letters to the right that correspond to the correct answers.

13. six and seventeen thousandths
   a. 6.017   b. $6\frac{17}{1000}$ c. 6.17   d. $6\frac{17}{100}$ [a] [b] [c] [d]

14. two hundred forty-six and three hundred-thousandths
   a. 246.0003   b. 246.00003   c. $246\frac{3}{100000}$ d. $246\frac{3}{10000}$ [a] [b] [c] [d]

Complete the following chart. The first row has been completed.

	Decimal	Fraction	Pronunciation
15.	.7	$\frac{7}{10}$	seven tenths
16.	.05		five hundredths
17.		$\frac{17}{100}$	
18.	.057		
19.			379 hundred-thousandths

1

READING AND WRITING DECIMALS

LESSON ONE: **Reading and Writing Decimal Fractions (Tenths)**

Instruction   Decimal fractions (sometimes simply called decimals) with only one digit to the right of the decimal point indicate **tenths**. They can be written as fractions with a denominator of 10.

Example

> .3 = 3 tenths = $\frac{3}{10}$
>
> .7 = 7 tenths = $\frac{7}{10}$

Exercise A   Write the fractional form of the following decimals.

1. .3 = ...   2. .4 = ...

3. .8 = ...   4. .7 = ...

5. .1 = ...   6. .6 = ...

7. .2 = ...   8. .1 = ...

9. .9 = ...   10. .5 = ...

11. .4 = ...   12. .8 = ...

13. .5 = ...   14. .3 = ...

15. .7 = ...   16. .2 = ...

Exercise B   Write these decimals as fractions.

1. .7 = ...   2. .3 = ...

3. .2 = ...   4. .1 = ...

5. .5 = ...   6. .9 = ...

7. .6 = ...   8. .4 = ...

9. .8 = ...   10. .2 = ...

LESSON TWO: Reading and Writing Decimal Fractions (Hundredths)

Instruction Decimals with two digits to the right of the decimal point indicate **hundredths**. They have a denominator of 100 when written as fractions.

Example

$.30 = 30$ hundredths $= \frac{30}{100}$

$.03 = 3$ hundredths $= \frac{3}{100}$

$.45 = 45$ hundredths $= \frac{45}{100}$

Exercise A Write the fractional form of the following decimals.

1. .25 = _____
2. .16 = _____
3. .01 = _____
4. .08 = _____
5. .60 = _____
6. .20 = _____
7. .09 = _____
8. .02 = _____
9. .33 = _____
10. .85 = _____
11. .07 = _____
12. .50 = _____
13. .39 = _____
14. .97 = _____
15. .11 = _____
16. .75 = _____
17. .44 = _____
18. .51 = _____
19. .23 = _____
20. .70 = _____

Exercise B Write these decimals as fractions.

1. $.\overline{34}$ = _____
2. .71 = _____
3. .06 = _____
4. .56 = _____
5. .33 = _____
6. .40 = _____
7. .75 = _____
8. .52 = _____
9. .16 = _____
10. .08 = _____
11. .98 = _____
12. .45 = _____

3

LESSON THREE: Reading and Writing Decimal Fractions (Thousandths)

Instruction   Decimals with three digits to the right of the decimal point indicate **thousandths**. They have a denominator of 1,000 when written as fractions.

Example

.435 = 435 thousandths = $\frac{435}{1000}$

.005 = 5 thousandths = $\frac{5}{1000}$

.400 = 400 thousandths = $\frac{400}{1000}$

Exercise A   Write the fractional form of the following decimals.

1. .006 = _____   2. .008 = _____

3. .238 = _____   4. .562 = _____

5. .600 = _____   6. .200 = _____

7. .025 = _____   8. .040 = _____

9. .070 = _____   10. .090 = _____

11. .186 = _____   12. .075 = _____

13. .217 = _____   14. .050 = _____

15. .101 = _____   16. .868 = _____

17. .225 = _____   18. .458 = _____

Exercise B   Write these decimals in fractional form.

1. .101 = _____   2. .030 = _____

3. .683 = _____   4. .135 = _____

5. .004 = _____   6. .602 = _____

7. .500 = _____   8. .984 = _____

9. .078 = _____   10. .002 = _____

11. .410 = _____   12. .930 = _____

13. .705 = _____   14. .706 = _____

LESSON FOUR:

Reading and Writing Decimal Fractions (Ten-thousandths)

Instruction   Decimals with four digits to the right of the decimal point indicate **ten-thousandths**. They have a denominator of 10,000.

Example

$.2311 = 2{,}311$ ten-thousandths $= \frac{2311}{10000}$

$.5000 = 5{,}000$ ten-thousandths $= \frac{5000}{10000}$

$.0004 = 4$ ten-thousandths $= \frac{4}{10000}$

$.0600 = 600$ ten-thousandths $= \frac{600}{10000}$

$.0020 = 20$ ten-thousandths $= \frac{20}{10000}$

Exercise A   Write the fractional form of the following decimals.

1. .0008 = ...

2. .0030 = ...

3. .0033 = ...

4. .0002 = ...

5. .0445 = ...

6. .0635 = ...

7. .2643 = ...

8. .3030 = ...

9. .5002 = ...

10. .4006 = ...

11. .0234 = ...

12. .5601 = ...

13. .5410 = ...

14. .7154 = ...

Exercise B   Write the fractional form of the following decimals.

1. .2304 = ...

2. .3333 = ...

3. .0613 = ...

4. .6011 = ...

5. .0001 = ...

6. .4432 = ...

7. .5306 = ...

8. .0106 = ...

9. .0103 = ...

10. .0010 = ...

11. .1011 = ...

12. .5050 = ...

13. .2410 = ...

14. .6162 = ...

5

LESSON FIVE: Reading and Writing Decimal Fractions (Hundred-thousandths)

Instruction

Decimals with five digits to the right of the decimal point indicate **hundred-thousandths**. They have a denominator of 100,000 when written as fractions.

Example

.00008 = 8 hundred-thousandths = $\frac{8}{100000}$

.00023 = 23 hundred-thousandths = $\frac{23}{100000}$

.05406 = 5,406 hundred-thousandths = $\frac{5406}{100000}$

.20355 = 20,355 hundred-thousandths = $\frac{20355}{100000}$

Exercise A

Write the fractional form of the following decimals.

1. .00044 = ...
2. .00018 = ...
3. .40003 = ...
4. .60002 = ...
5. .00200 = ...
6. .63000 = ...
7. .01023 = ...
8. .00408 = ...
9. .00002 = ...
10. .00009 = ...
11. .04014 = ...
12. .00142 = ...
13. .72710 = ...
14. .36891 = ...
15. .80345 = ...
16. .00005 = ...

Exercise B

Write these decimals in fractional form.

1. .52301 = ...
2. .00100 = ...
3. .06304 = ...
4. .00009 = ...
5. .32789 = ...
6. .10019 = ...
7. .00001 = ...
8. .21006 = ...
9. .10006 = ...
10. .98765 = ...
11. .01023 = ...
12. .12345 = ...
13. .65423 = ...
14. .00242 = ...

Reading and Writing Decimals

Instruction   Decimals are read according to the number of digits to the right of the decimal point.

Example

hundred-thousandths

ten-thousandths

thousandths

hundredths

tenths

decimal point ⟶ .   0   0   0   0   0

Exercise A   Write each of the following in decimal form and fractional form.

1. 8 ten-thousandths ....

2. 23 thousandths ....

3. 4 tenths ....

4. 2,038 hundred-thousandths ....

5. 55 hundredths ....

6. 62 thousandths ....

7. 8,324 ten-thousandths ....

Exercise B   Write each of the following in decimal form and fractional form.

1. 68 ten-thousandths ....

2. 5 tenths ....

3. 56 thousandths ....

4. 12,038 hundred-thousandths ....

5. 4 thousandths ....

6. 1,036 hundred-thousandths ....

7. 433 ten-thousandths ....

7

Reading and Writing Mixed Decimals

Instruction

The digits to the left of the decimal point indicate a whole number. When there is a whole number mixed with a decimal fraction, the decimal point is read "and."

Example

43.2 = forty-three **and** two tenths = $43\frac{2}{10}$

4,006.23 = four thousand six **and** twenty-three hundredths = $4,006\frac{23}{100}$

300.006 = three hundred **and** six thousandths = $300\frac{6}{1000}$

Exercise A   Write the decimal form and fractional form of each of the following.

1. four and twenty-three hundredths

--   --

2. one hundred sixty-three and eight thousandths

--   --

3. five thousand eight and two thousand four hundred sixty-three ten-thousandths

--   --

4. forty-four and four hundred-thousandths

--   --

5. twenty thousand three and twenty thousandths

--   --

Exercise B   Write the decimal form and fractional form of each of the following decimals.

1. two hundred sixty-eight and four hundredths

--   --

2. ten and one thousand two ten-thousandths

--   --

3. ninety-four and seventy-six thousandths

--   --

4. five hundred and two tenths

--   --

UNIT 1—REVIEW

Instruction Sometimes, you hear people say "four point two five" when they see a decimal like 4.25. However, you should be able to read decimals according to tenths, hundredths, etc. If you read decimals properly, it will be easier to write them in fractional form.

Exercise A Complete the following chart. The first row has been completed.

	Decimal	Fraction	Pronunciation
1.	.03	$\frac{3}{100}$	3 hundredths
2.	.4		4 tenths
3.		$\frac{5}{100}$	
4.	.023		
5.			426 hundred-thousandths
6.	.00024		
7.		$\frac{1045}{100000}$	
8.			530 ten-thousandths
9.	.004		
10.		$\frac{802}{1000}$	
11.			16 thousandths
12.	.141		
13.		$\frac{84}{100}$	
14.			987 hundred-thousandths
15.		$\frac{7}{10}$	
16.	.5		
17.			21 hundredths
18.		$\frac{666}{1000}$	
19.	.019		
20.			702 ten-thousandths
21.			91 thousandths

9

Exercise B   Complete the following chart. The first one has been done.

	Decimal	Fraction	Pronunciation
1.	.048	$\frac{48}{1000}$	48 thousandths
2.		$\frac{7}{10}$	
3.			4 hundredths
4.	.0006		
5.		$16\frac{2}{1000}$	
6.			12 and 3 hundredths
7.	436.03		
8.		$213\frac{25}{10000}$	
9.			23 and 23 thousandths
10.	1,833.018		
11.		$\frac{49}{100}$	
12.			77 and 6 tenths
13.	240.013		
14.		$66\frac{433}{1000}$	
15.			20 and 11 ten-thousandths
16.	.0076		
17.		$39\frac{62}{100}$	
18.			18 and 16 hundredths
19.	.45		
20.		$2,465\frac{9870}{10000}$	

STATE EMPLOYMENT OFFICE

Every state has a department that deals with industry and labor. The name of the department may vary from state to state. Usually, the department is broken up into several divisions. One of the main divisions of the labor department will be the state employment service.

Each state has a special employment office. Its name varies, but it is usually called the state employment commission or state employment office.

The state employment offices offer many services through their local branches. Your local state employment office can—

- help you find a job that is right for you.
- give you employment counseling to improve your chances of finding a job you are qualified for.
- give you employment tests to find out what you can do best.
- refer you to training programs to learn new skills.
- refer you to other agencies for financial aid.

State employment agency services are free. If you go to a private employment agency, either you or your employer will have to pay a fee.

You do not need to have an appointment to visit your local state employment office, but you will need to know the hours that the office is open. When you go, take the following things with you,

- your social security card, if you have one.
- a list of the companies you have worked for and their addresses.
- proof of military service, if you are a veteran. (There may be special services available for veterans.)

When you get to the employment office, go to the information desk. Tell the person that you are looking for a job. You will be asked to fill out an application form. You will then talk to an employment counselor who will try to help you find a job.

To find your nearest state employment office, look in the white pages of the phone directory under the name of the state you live in. For instance, if you live in Missouri, look under MISSOURI—STATE OF. All of the local state offices will be listed under this heading. Find the listing for the employment office and write down the phone number and address. Call to find out what their business hours are. Record the information below.

Nearest State Employment Office

Address: ....

....

Phone Number: ....

Hours: ....

UNIT 2—PRETEST
COMPARING DECIMALS

Find the mixed decimal and the mixed fraction of each. Blacken the letters to the right that correspond to the correct answers.

1. five dollars and eighty-four cents
   a. $5.84   b. $58.40   c. $5\frac{84}{100}$ d. $5\frac{84}{1000}$ [a] [b] [c] [d]

2. forty-six dollars and seventy-five cents
   a. $460.75   b. $46.75   c. $46\frac{75}{1000}$ d. $46\frac{75}{100}$ [a] [b] [c] [d]

Write *true* after each true statement. Write *false* after each false statement.

3. 11. = 11.00   ----------------   4. .07 = .070   ----------------

5. .063 = .06300   ----------------   6. 6.58 = 6.58000   ----------------

Determine whether or not the decimals are equal. Write *equal* or *not equal* in the blanks.

7. .9
   .90   ----------------   8. .04
                                    .40   ----------------

9. .067
   .0067   ----------------   10. .36
                                      .3600   ----------------

Circle the decimal in each pair which has the greater value.

11. .9 or .092   12. .266 or .4   13. .149 or .14

Write each decimal as a fraction and as a fraction reduced to lowest terms.

14. .10   ----------------   ----------------

15. 4.08   ----------------   ----------------

Write the decimal with the greater value in the blank. If the decimals have equal values, write *equal* in the blank.

16. .11, .127   ----------------

17. 16.7, 1.67   ----------------

COMPARING DECIMALS

LESSON ONE: **Writing a Whole Number and a Decimal Fraction**

Instruction

The digits to the left of the decimal point indicate a whole number. Digits to the right of the decimal point indicate a fractional part of the whole number 1.

Example

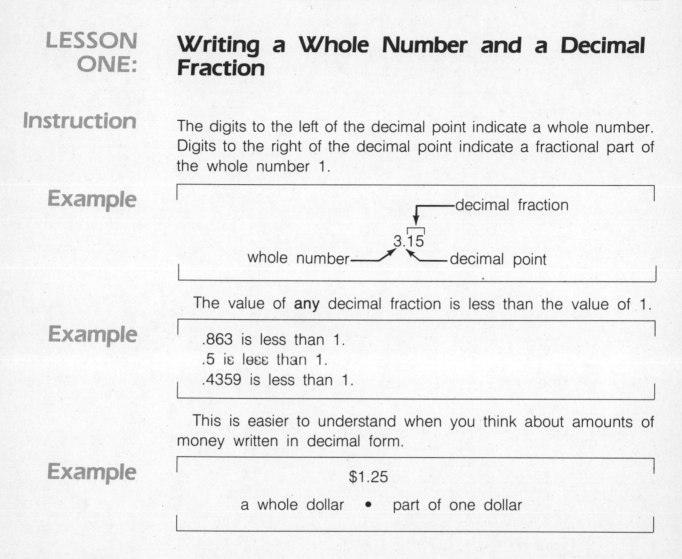

decimal fraction

3.15

whole number —— decimal point

The value of **any** decimal fraction is less than the value of 1.

Example

.863 is less than 1.
.5 is less than 1.
.4359 is less than 1.

This is easier to understand when you think about amounts of money written in decimal form.

Example

$1.25

a whole dollar • part of one dollar

Directions

Complete the following chart. The first one has been done.

	Words	Mixed Decimal	Mixed Fraction
1.	one dollar and sixty cents	$1.60	$1\frac{60}{100}$
2.	six dollars and three cents		
3.	twenty dollars and forty cents		
4.	one hundred eight dollars and twenty-five cents		
5.	four dollars and five cents		
6.	one dollar and ninety-nine cents		
7.	seven dollars and seventeen cents		

**Affixing Zeros to the Right of a Decimal
Point**

Instruction Zeros may be added to the **right** of a decimal point without
changing the value of the number.

Example

$6 = $6.00
2.4 = 2.40 = 2.400

Exercise A Determine whether each of the following statements is true or
false. Write *true* or *false* in each blank.

1. 5. = 5.0 ....

2. 2.34 = 2.340 ....

3. .063 = .063000 ....

4. .7 = .7000 ....

5. .08 = .080 ....

6. .4 = .4000 ....

7. .07 = .070 ....

8. .69 = .69000 ....

9. 5.74 = 5.740 ....

Exercise B Write *true* in the blanks beside the statements which are true.
Write *false* in the blanks beside those that are false.

1. 2.47 = 2.4700 ....

2. 10. = 10.00 ....

3. .08 = .080 ....

4. .05 = .050 ....

5. .020 = .0200 ....

6. 60.62 = 60.620 ....

7. 9. = 9.000 ....

8. 3.52 = 3.5200 ....

"STRETCH-JERSEY"
● (MACHINE-WASHABLE)
(TOSS IN DRYER)
100 PERCENT COTTON KNIT

SALE!
$3.98

Dropping Zeros at the End of a Decimal

Instruction

Zeros at the end of a decimal have **no value** and may be dropped without any change in the value of the number.

Example

$$.5000 = \frac{5000}{10000} = \frac{5}{10} = \frac{1}{2}$$

$$.500 = \frac{500}{1000} = \frac{5}{10} = \frac{1}{2}$$

$$.50 = \frac{50}{100} = \frac{5}{10} = \frac{1}{2}$$

$$.5 = \frac{5}{10} = \frac{1}{2}$$

However, zeros between the decimal point and a digit do have value and may not be ignored.

Example

$$.5 = \frac{5}{10} = \frac{1}{2}$$

$$.05 = \frac{5}{100} = \frac{1}{20}$$

Therefore, .5 does not equal .05.

Exercise A

Write each decimal as a fraction and reduce the fraction to lowest terms. Determine whether or not the decimals are equal. Write *equal* or *not equal* in the blanks.

1. .8
   .80   _____

2. .25
   .250   _____

3. .02
   .20   _____

Exercise B

Write each decimal as a fraction and reduce the fraction to lowest terms. Determine whether or not the decimals are equal. Write *equal* or *not equal* in the blanks.

1. .6
   .60   _____

2. .08
   .80   _____

3. .002
   .0020   _____

Comparing Decimals

Instruction

When comparing the value of decimals, it is easier to do if they have the same number of decimal places. Decimals can be made to have the same number of decimal places by adding or dropping zeros.

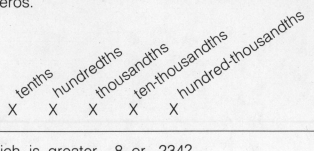

Example

Which is greater, .8 or .234?

.234 ⟶ .234 ⟶ .800 (.8) is greater.
.8 = .800 .800

.234 has three decimal places. .8 has only one. You can add two zeros to the right of .8 so that it will have the same number of decimal places as .234. Then compare the decimals. .800 is greater than .234. Therefore, .8 is greater than .234.

Exercise A

Determine which decimal has the greater value. Circle it.

1. .3 or .65
2. .9 or .130
3. .025 or .12
4. .4 or .53
5. .77 or .092
6. .54 or .536
7. .2 or .80
8. .013 or .13

Exercise B

Circle the decimal in each pair which has the greater value.

1. .5 or .236
2. .03 or .2
3. .186 or .32
4. .6 or .06
5. .1 or .099

Writing Whole Numbers with a Decimal Point

Instruction

Every whole number has a decimal point, but you usually never see it. You can put a decimal point after the right-hand digit of any whole number.

You can also put zeros to the right of the decimal point of any whole number without changing its value.

Example

$8 = $8. = $8.00
$16 = $16. = $16.00
250 = 250. = 250.000

Exercise A

Complete the following chart. The first one has already been done.

	Whole Number	Whole Number and Decimal Point	One Zero	Two Zeros
1.	8	8.	8.0	8.00
2.			25.0	
3.		633.		
4.	400			
5.				1.00
6.	75			
7.		36.		

Exercise B

Complete the chart.

	Whole Number	Whole Number and Decimal Point	One Zero	Two Zeros
1.			44.0	
2.	750			
3.		2.		
4.				13.00
5.	10			
6.	234			

17

LESSON SIX:

Converting a Decimal to a Fraction and Reducing

Instruction

A fraction is **reduced** by dividing the numerator and the denominator by the **same number**. When changing a decimal fraction to its fractional equivalent, it usually is best to reduce the fraction to lowest terms when possible.

Example

$$.25 = \frac{25}{100} = \frac{1}{4}$$

$$1.8 = 1\frac{8}{10} = 1\frac{4}{5}$$

$$15.4 = 15\frac{4}{10} = 15\frac{2}{5}$$

Exercise A

Complete the following chart. The first one has been done.

	Decimal	Fraction	Lowest Terms
1.	1.06	$1\frac{6}{100}$	$1\frac{3}{50}$
2.	.5		
3.	20.04		
4.	1.6		
5.	18.2		
6.	123.025		
7.	4.25		

Exercise B

Complete the following chart.

	Decimal	Fraction	Lowest Terms
1.	.15		
2.	53.80		
3.	.40		
4.	4.2		
5.	66.666		
6.	74.82		

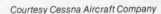

Courtesy Cessna Aircraft Company

18

UNIT 2—REVIEW

Exercise A   Determine whether each of the following statements is true or false. Write *true* or *false* in each blank.

1. $2.4 = 2\frac{2}{5}$ ------------------

2. $18 = 18.00$ ------------------

3. $16.3 = 16.30$ ------------------

4. $.03 = .3$ ------------------

5. $.5000 = .5$ ------------------

6. $.15 = \frac{3}{20}$ ------------------

7. $14 = 1.4$ ------------------

8. $7.05 = 7\frac{1}{20}$ ------------------

9. $100 = 100.0$ ------------------

10. $.008 = .08$ ------------------

Exercise B   Determine which decimal in each pair has the greater value and write it in the blank. If the decimals have equal values, write *equal* in the blank.

1. .4, .43   ------------------

2. .008, .08   ------------------

3. 14, .99   ------------------

4. 2.6, 1.9   ------------------

5. 3.18, 3.1   ------------------

6. .230, .2300   ------------------

7. 5, 5.00   ------------------

8. .127, .9   ------------------

9. .45, .438   ------------------

10. .2, .002   ------------------

11. .75, .649   ------------------

Exercise C

Determine which number in each pair has the greater value and write it in the blank. If the values are equal, write *equal*.

1. 2.6, 2.600 _____

2. 1.4, 14 _____

3. 3.25, $3\frac{1}{4}$ _____

4. .8, .008 _____

5. $22\frac{1}{2}$, 22.5 _____

6. 80, 80. _____

7. 129.4, $129\frac{2}{5}$ _____

8. .15, .1500 _____

9. 18.04, $18\frac{4}{10}$ _____

10. 630.30, 630.3 _____

11. 46.041, 46.0410 _____

12. .03, .3 _____

13. 18, 1.8 _____

14. 146.03, 146.030 _____

15. .006, .060 _____

16. .75, .756 _____

17. $51\frac{2}{5}$, 51.4 _____

18. 8, 8.0 _____

19. .16, 0.16 _____

20. .95, $\frac{95}{100}$ _____

21. 66.2, 6.62 _____

22. .77, $\frac{77}{1000}$ _____

23. 16.1, .1610 _____

24. $6\frac{4}{5}$, 6.8 _____

25. .75, $\frac{4}{5}$ _____

26. 202.2, 20.22 _____

20

KEEPING FAMILY RECORDS

Charlotte Green had been working for a year. This was going to be the first time she would file an income tax return. She wanted to do it herself, but she hadn't kept any receipts, canceled checks, or other records. She had little upon which she could base any itemized deductions.

Homer Mabry thinks he has been caught in a credit trap. He needs to look at his copy of the credit contract, but he can't find it.

Monica Mills needed a copy of her birth certificate. She was applying for social security benefits. She looked everywhere but could not find it. Finally, she had to write her state health department for a copy.

All of these people needed their important papers and records. But none of them had kept the records and papers in a special, safe place.

Everyone should keep important family records and papers, such as receipts and canceled checks, credit contracts, birth certificates, wills, insurance policies, real estate deeds, etc.

Keeping records in a safe place will protect them against loss. It will also save you time when you need to find them. Some records can be safely kept in a fireproof box in your home. (You can keep them in a cardboard box, but they will not be protected from fire.) Other records should be kept in a safe-deposit box at a bank or savings and loan. There is usually a small fee charged for each safe-deposit box you rent.

The chart below shows what records and papers should be kept. The best place to keep each record is marked with an *X*. The chart also indicates how long you should keep each record. You may not have all of the records shown in the chart. And there may be other records that you will want to keep in addition to those listed.

Where To Keep Family Records

Type of Record	Where To Keep It		Keep How Long
	Home File	Safe-Deposit Box	
adoption papers		X	for life
appliance instruction books	X		during ownership
automobile titles		X	during ownership
bank account books	X		while open
bank monthly statement	X		2 years
bills of sale	X		6 years
birth certificates		X	for life
canceled checks (general)	X		2 years
canceled checks (major transactions)	X		5 years
citizenship papers		X	for life
contracts		X	while in force + 3 years
educational records		X	for life
employment (work) records	X		life (optional)
employment benefit records	X		while employed
family business records	X		6 years

Type of Record	Where To Keep It		Keep How Long
	Home File	Safe-Deposit Box	
household property inventory	X		revise yearly
income records, pay statements	X		5 years
insurance policies		X	while in force
investments (stocks and bonds)		X	during ownership
leases	X		while in force
marriage and divorce certificates		X	for life
real estate deeds		X	during ownership + 6 years
social security cards	X		for life
tax records (income, property, sales)	X		5 years
U.S. savings bonds		X	cash-in date
warranties and guarantees	X		while in force
wills		X	for life

UNIT 3—PRETEST
ADDING DECIMALS

Add the following decimals.

1.   2 4 2.0 5 6
     + 6 7.8 3 1

2.   1 2 3.4 6
     + 4 2 5.5 7

3.   9,8 5 1.6 8
     + 4,3 2 3.4 2

4.   8 0 0.0 6 6
     + 2 4 3.8 9 7

Set up the following problems and add.

5. .68 + 32.161 + .5 =
6. 4.015 + 5,041.3 + .85 =
7. 6.8352 + 14.057 + .81 =
8. 186.204 + 5.2103 + 222.97 =

Write these problems in column form and add.

9. 12.341 + 16 + .07 =
10. $18 + $6.70 + $.17 =
11. $55.00 + $40 + $.66 =
12. 75.003 + 64 + 24.2 =

Solve the following word problems. Blacken the letter to the right that corresponds to the letter before the correct answer.

13. Cornell spent $46.00 on groceries, $25.63 for a birthday present, and $9.82 for dry cleaning. How much did she spend in all?
    a. $82.50   b. $81.45   c. $81.05   d. $80.05   a b c d

14. Nelda bought two gallons of paint costing $15.99 a can, a brush which cost $1.79, and a pan and roller set for $3.99. How much did she pay for paint supplies?
    a. $27.17   b. $27.07   c. $21.17   d. $37.76   a b c d

15. When Teodoro bought his used car, the odometer showed 12,686.4 miles. In one year he drove it 13,294 miles. What did the odometer read then?
    a. 29,580   b. 25,980.4   c. 25,990   d. 26,980   a b c d

ADDING DECIMALS

Placing a Decimal Point in Sums

Instruction   When decimals are added, the decimal point must be placed in the answer. The decimal point in the answer **must line up with the decimal point in the addends.** (When money is added, you also need to place a dollar sign in front of the answer.)

Example

```
  23.14  ──►    23.14  ──►    23.14
+ 15.06       + 15.06       + 15.06
               3820          38.20
```

```
$1.43  ──►   $1.43  ──►   $1.43
+ .25       + .25        + .25
             168         $1.68
```

Exercise A   Add the following decimals.

1.   162.134
   + 14.052

2.   .241
   + .618

3. $6.00
   + .75

4. $24.20
   + 1.65

5. $126.04
   + 3.84

6. $21.68
   + 9.07

7.   4.1414
   + .0333

8. $523.60
   + 76.50

9.   4.6323
   + .0109

Exercise B   Add the following decimals.

1.   386.42
   + 12.08

2.   .302
   + .157

3. $8.00
   + .66

4. $53.68
   + 1.25

5. $125.03
   + 4.16

6.   681.042
   + 397.107

23

LESSON TWO: Aligning Decimal Points When Adding

Instruction   When adding decimals, line up the decimal points one on top of the other.

Example

$$23.06 + 4.8 + 104.21 \longrightarrow$$

```
    2 3.0 6
       4.8
 + 1 0 4.2 1
  1 3 2.0 7
```

$$\$21.50 + \$.25 + \$.05 \longrightarrow$$

```
  $ 2 1.5 0
       .2 5
  +    .0 5
  $ 2 1.8 0
```

Exercise A   Set up the following problems and add.

1. 1,405.21 + 3.008 + .33 =   2. 1.045 + 18.4 + .02 =

3. $1.05 + $100.25 + $23.40 =   4. $.75 + $.20 + $.09 =

Exercise B   Set up the following problems and add.

1. 1.61 + 120.13 + .4 =   2. .006 + 23.04 + .01 =

3. $.44 + $1.15 + $2.60 =   4. $.36 + $.21 + $.85 =

Courtesy Georgia-Pacific Corporation

LESSON THREE: Adding a Whole Number and a Decimal Fraction

Instruction   When adding a whole number and a decimal fraction, be careful to line up the decimal points. Remember, whole numbers should be on the left-hand side of the decimal point.

Example

$$148.03 + 6 + .82 \longrightarrow \begin{array}{r} 148.03 \\ 6.00 \\ +\ .82 \\ \hline 154.85 \end{array}$$

$$\$.55 + \$4 + \$3.60 + \$2 \longrightarrow \begin{array}{r} \$\ \ .55 \\ 4.00 \\ 3.60 \\ +\ 2.00 \\ \hline \$10.15 \end{array}$$

Exercise A   Write these problems in column form and add.

1. 41.008 + 23 + .79 =

2. 312 + .006 + 46.7 =

3. $1.64 + $8 + $.30 =

4. $12 + $.04 + $1.80 =

Exercise B   Set up these problems and add.

1. 60.003 + 410 + 3.06 =

2. 39 + 00.04 + 20.008 =

3. $13.07 + $59 + $.30 =

4. $34 + $.10 + $21.06 =

5. 264.2 + 36 + .003 =

6. $19.87 + $.56 + $113 =

LESSON FOUR:
Solving Word Problems Involving the Addition of Decimals

Directions   Solve the following word problems.

1. Mrs. Osagede paid some of her bills yesterday. She wrote these checks: $375.00 for rent, $225.00 for car loan, $168.75 for utilities, and $39.18 for telephone. What was the total amount she paid out?

   ----

2. A salesperson keeping record of her mileage logged the following distances: 25.3 miles on Monday, 16.8 miles on Tuesday, 34.4 miles on Wednesday, and 20 miles on Thursday. How many total miles did she travel on those days?

   ----

3. When Sheldon Markowski bought a new washing machine, he paid $55 down and then paid $45.78 each month for seven months. How much did his washer cost?

   ----

4. Jesse Escamilla's take-home pay is $1,136.51 a month. $97.71 is deducted for social security; $224.12 is deducted for federal income tax. What is his monthly gross pay? (Gross pay is the grand total before any deductions are made.)

   ----

5. A road construction crew finished surfacing 5.7 miles of highway one week and 4.3 miles the next week. How many miles did they complete in two weeks?

   ----

6. Milton Lindell is figuring the monthly grocery bill. How much does the family spend on groceries if they pay $65.47 the first week, $72.16 the second week, $57.18 the third week, and $50 the fourth week?

   ----

7. Vikki bought a case of sodas at $6.80 a case plus a $2.20 bottle and case deposit. What was her total bill?

UNIT 3—REVIEW

Exercise A   Write these problems in column form and add.

1. 1.003 + 141.2 + .47 =   **2.** 12 + .006 + 128 =

3. 5.6 + 3 + .1829 =   **4.** .391 + 80 + 4.63 =

5. .001 + .150 + .12 =   **6.** 6.341 + .02 + 1.6 =

7. .9764 + 147 + 67.31 =   **8.** 53.261 + 3.59 + .0068 =

9. 16.35 + 211.684 + 75.0697 =

10. 321.6 + 485.791 + .0001 =

Courtesy Allied Chemical Corporation

Exercise B   Write these problems in column form and add.

1. .16 + .04 + .33 =

2. 213.12 + 141.008 + 16.013 =

3. $1.48 + $22 + $.15 =

4. $180 + $.05 + $.30 =

5. $16.50 + $7 + $.98 =

6. 89.683 + 35.02 + 19.1 =

7. 64.1 + 18 + 49.003 =

8. 47.91 + 5.3 + .00163 =

9. .22 + 46.84 + 95.6 =

10. .042 + 0.18 + 19.1 =

11. .49 + 2.62 + 4.903 =

12. 801 + 148.06 + 94.306 =

13. 19.74 + 81.02 + .10603 =

14. 84.64 + .223 + 5,061.2 =

CHECKBOOK RECORDS

Paying for things by check has advantages over paying with cash. By using a checking account, you can be sure that your money will not be lost or stolen. When you pay by check, the canceled check is proof of payment.

Checking accounts are convenient and safe. You should keep up-to-date and accurate checkbook records which include amounts of deposits made and checks written on the account. Checkbooks provide a space for record keeping. Here is a typical record of transactions.

Check No.	Date	Check Issued To	(−) Amount of Check	(+) Amount of Deposit	Balance 143.16
	7-31			1,355.62	1,498.78
354	8-1	Home Finance Co.	820.23		678.55
355	8-1	City Utility	124.16		554.39
	8-3			115.00	669.39
356	8-4	Bell Telephone	29.80		639.59
357	8-4	Grocery Mart	88.54		551.05
358	8-4	Cash	70.00		481.05

Directions

This is another sample checkbook register. Study the entries and then answer the questions.

Check No.	Date	Check Issued To	(−) Amount of Check	(+) Amount of Deposit	Balance 324.18
461	9-14	Quick Food Store	47.62		
	9-15			289.14	
462	9-16	American Ins. Co.	24.89		
463	9-16	Hayes Furniture	116.14		
464	9-16	Cash	15.00		
465	9-17	Self-Serv Gas	8.60		
	9-18			44.00	
466	9-19	Acme Finance Co.	86.43		

1. What is the balance at the end of the day on September 15? _____

2. What was the total amount of checks written on 9-16? _____

UNIT 4—PRETEST
SUBTRACTING DECIMALS

Subtract the following decimals.

1. $\begin{array}{r} 2.5520 \\ -1.6726 \\ \hline \end{array}$ 2. $\begin{array}{r} \$643.65 \\ -354.76 \\ \hline \end{array}$

3. $\begin{array}{r} \$4.83 \\ -3.79 \\ \hline \end{array}$ 4. $\begin{array}{r} 754.716 \\ -599.092 \\ \hline \end{array}$

Set up the following problems and subtract.

5. $795.12 − $.97 =
6. 416.060 − 8.9 =
7. $9.98 − $6.99 =
8. 2.0163 − 2.0079 =

Rewrite and solve the following problems.

9. 64 − 12.88 =
10. $75 − $69.95 =
11. 7 − 5.0684 =
12. $90 − $80.63 =

Set up these problems and subtract.

13. 638.6 − .659 =
14. $5 − $3.85 =
15. $57.03 − $48.97 =
16. 78.0939 − 9.4 =

Solve the following word problems. Blacken the letter to the right that corresponds to the letter before the correct answer.

17. With his allowance of $8.00 Garth bought his lunch at school two times at a cost of $1.10 per lunch. He put $5.00 in his savings account. How much money does Garth have left to spend?
    a. $.080   b. $.80   c. $80   d. $8.00

SUBTRACTING DECIMALS

LESSON ONE: Placing a Decimal Point in Differences

Instruction   When decimal fractions are subtracted, a decimal point must be placed in the answer. The decimal point in the answer has to line up with the decimal points in the minuend and subtrahend.

Example

$$
\begin{array}{r} 1\,5.4\,8 \\ -\,3.2\,1 \\ \hline \end{array}
\longrightarrow
\begin{array}{r} 1\,5.4\,8 \\ -\,3.2\,1 \\ \hline 1\,2\,2\,7 \end{array}
\longrightarrow
\begin{array}{r} 1\,5.4\,8 \\ -\,3.2\,1 \\ \hline 1\,2.2\,7 \end{array}
$$

Remember, when amounts of money are figured, place the dollar sign to the left of the answer.

$$
\begin{array}{r} \$\,4\,6.5\,5 \\ -\,2.3\,4 \\ \hline \end{array}
\longrightarrow
\begin{array}{r} \$\,4\,6.5\,5 \\ -\,2.3\,4 \\ \hline 4\,4\,2\,1 \end{array}
\longrightarrow
\begin{array}{r} \$\,4\,6.5\,5 \\ -\,2.3\,4 \\ \hline \$\,4\,4.2\,1 \end{array}
$$

Exercise A   Subtract.

1. $\begin{array}{r} 1\,2.8\,4\,5 \\ -\,1.6\,2\,4 \\ \hline \end{array}$

2. $\begin{array}{r} 3.2\,7\,1\,5 \\ -\,.1\,9\,8\,2 \\ \hline \end{array}$

3. $\begin{array}{r} \$\,7\,8.0\,9 \\ -\,2\,5.1\,0 \\ \hline \end{array}$

4. $\begin{array}{r} \$\,1\,2\,7.1\,5 \\ -\,3\,2.7\,5 \\ \hline \end{array}$

5. $\begin{array}{r} \$\,5.6\,5 \\ -\,.4\,0 \\ \hline \end{array}$

6. $\begin{array}{r} 8\,4.6\,1\,3 \\ -\,5\,9.9\,0\,7 \\ \hline \end{array}$

Exercise B   Subtract.

1. $\begin{array}{r} 2,1\,7\,2.8 \\ -\,9\,3\,2.3 \\ \hline \end{array}$

2. $\begin{array}{r} 2.2\,5\,2\,0 \\ -\,.1\,5\,0\,6 \\ \hline \end{array}$

3. $\begin{array}{r} \$\,2\,1\,8.4\,5 \\ -\,9\,2.2\,8 \\ \hline \end{array}$

4. $\begin{array}{r} \$\,2\,7.1\,8 \\ -\,4.2\,5 \\ \hline \end{array}$

5. $\begin{array}{r} \$\,4.8\,3 \\ -\,.5\,5 \\ \hline \end{array}$

6. $\begin{array}{r} 5\,8.0\,0\,6 \\ -\,3\,9.7\,5\,9 \\ \hline \end{array}$

LESSON TWO: Aligning Decimal Points When Subtracting

Instruction When subtracting decimals, line up the decimal points one on top of the other.

Example

$$63.87 - 2.6 \longrightarrow \begin{array}{r} 63.87 \\ -\ 2.6 \\ \hline 61.27 \end{array}$$

$$\$12.65 - \$.45 \longrightarrow \begin{array}{r} \$12.65 \\ -\ .45 \\ \hline \$12.20 \end{array}$$

Exercise A Set up the following problems and subtract.

1. 21.060 − 3.006 =   2. 42.973 − 3.2 =

3. $586.12 − $24.08 =   4. $9.80 − $.75 =

Exercise B Set up the following problems and subtract.

1. 14.603 − 5.1 =   2. 8.36 − .21 =

3. $436.42 − $14.10 =

4. $22.63 − $7.14 =

Subtracting a Decimal from a Whole Number

Instruction

When subtracting a decimal from a whole number, be sure the whole number is on the left-hand side of the decimal. If borrowing is necessary, place zeros to the right of the decimal and then borrow.

Example

$$24 - 1.385 \longrightarrow \begin{array}{r} \overset{\overset{9\ 9}{3\ 10\ 10\ 10}}{2\,4.0\,0\,0} \\ -\ 1.3\,8\,5 \\ \hline 2\,2.6\,1\,5 \end{array}$$

$$\$6 - \$4.25 \longrightarrow \begin{array}{r} \$\overset{\overset{9}{5\ 10\ 10}}{6.0\,0} \\ -\ 4.2\,5 \\ \hline \$\,1.7\,5 \end{array}$$

Exercise A Rewrite and solve the following problems.

1. 52 − 14.168 =

2. 3 − .262 =

3. $1 − $.35 =

4. $89 − $88.50 =

5. $16 − $4.43 =

6. 39 − 18.064 =

Exercise B Rewrite and solve the following problems.

1. 47 − 13.174 =

2. 9 − .6433 =

3. $2 − $.67 =

4. $74 − $70.74 =

LESSON FOUR:
Solving Word Problems Involving the Subtraction of Decimals

Directions   Solve the following word problems.

1. One automobile piston has a diameter of 4.003 inches. Another piston has a diameter of 3.821 inches. How much larger is the first piston?

    ----

2. Oscar's car averages 21.3 miles per gallon of gasoline. Karlyn's car gets 16.5 miles per gallon. Oscar's car averages how many more miles per gallon?

    ----

3. Ursula bought a lamp on sale for $16.85. The original price was $20. How much did she save?

    ----

4. Armando bought a shirt for $12.99 and a tie for $3.75. The tax was $.84. How much change should he receive from $20?

    ----

5. Your checkbook shows a balance of $183.65 before you write a $24.62 check for groceries and a $6.50 check for medicine. What is your new balance?

    ----

6. A plumber needs to replace 11.5 feet of pipe in a home. She has 6.5 feet in her truck. How much additional pipe does she need?

    ----

7. One store advertises coffee for $3.19 a pound. Another store sells coffee at $2.99 per pound. What is the savings on coffee at the second store?

    ----

8. When buying a new television set, Lou paid $75 down in cash and charged the balance. How much more did he owe on the $319.95 set?

9. The Morgans' utility statement showed that they owed $89.86. This amount included a $5 charge for late payment. How much was the original bill?

LESSON FIVE: Solving Word Problems Involving the Addition and Subtraction of Decimals

Directions

Solve the following word problems.

1. Jeff Tait wants to buy an electric razor. The price in the store is $29.99. He can buy the same one from the store's catalog for $24.50. Shipping and handling charges for the catalog purchase amount to $3.78. Which is the better deal?

   ....

2. Judy Greenfield takes her car in for service. She has a coupon worth $1.00 on the first $6.50, $2.00 on the next $6.50, and $3.00 on the next $6.50. How much is her final bill if the service charges are $22.75 before she presents the coupon?

   ....

3. Bill Triplett had a balance in his checkbook of $249.78. He wrote a check for $24.32, deposited $75.89, and wrote another check for $46.57. What was his new balance?

   ....

4. Weatherperson Sandra Dickson reported that the barometric pressure today was 30.02. Yesterday she reported that it was 29.86. How much did the barometric pressure change?

   ....

5. Jim French has just finished his junior year of college. His cumulative grade point average is 1.83. In order to graduate he needs an average of 2.00. How much must he increase his average to be able to graduate?

   ....

UNIT 4—REVIEW

Rewrite in computational form and subtract.

1. $163.682 - 41.5 =$ **2.** $361.008 - 4.595 =$ **3.** $60 - $44.28 =$

4. $4 - $.44 =$ **5.** $260.06 - $3 =$ **6.** $193 - 37.06 =$

7. $89.0309 - 6.3 =$ **8.** $56.59 - $4.17 =$ **9.** $100 - $13.75 =$

10. $519.3 - 68.009 =$ **11.** $37.984 - 16 =$ **12.** $45.01 - $.37 =$

13. $616.4 - 80.34 =$ **14.** $62.489 - 4.444 =$ **15.** $39.173 - 6.073 =$

16. $16.38 - .569 =$ **17.** $86.90 - $14.12 =$ **18.** $541.01 - 5.956 =$

19. $268.316 - 51.4 =$ **20.** $87.6345 - 13.75 =$ **21.** $106.46 - $77.97 =$

22. $60.26 - $30.90 =$ **23.** $65 - $19.65 =$ **24.** $5 - $.85 =$

Exercise B   Rewrite in computational form and subtract.

1. 18.4 – 13.168 =

2. 2,633.56 – 21 =

3. $48.50 – $2.31 =

4. $12 – $4.69 =

5. $123.54 – $3 =

6. 117.3 – 89.4 =

7. $87.01 – $4.56 =

8. $25 – $10.99 =

9. 875.321 – 36.497 =

10. 324 – 126.993 =

11. $14.75 – $12.26 =

12. 181.65 – 13.3 =

13. 26.84 – 17.003 =

14. 35 – 1.0069 =

15. 23.4 – 18.26 =

16. 37 – 14.024 =

17. $675 – $27.56 =

18. 172 – 66.345 =

19. 15.4758 – 6.3275 =

20. $.27 – $.19 =

21. 81.56 – 33.1 =

22. 73 – 7.41 =

23. 575.123 – .0534 =

24. $36.71 – $17.87 =

EARNINGS STATEMENTS

Shown below is a typical earnings statement. An earnings statement should be included with each paycheck you receive. It will show the number of hours you worked, the pay earned, and the deductions made from your pay. **Gross pay**, or total earnings, is the total amount of money you earned before deductions. **Net pay** is equal to your gross pay minus all deductions.

Directions
Look at the earnings statement and then answer the questions.

EMPLOYEE NO.	REG. HRS.	O.T. HRS.						PAY PERIOD ENDING	
9102050	80	18						3-15-84	

CURRENT PAY PERIOD

REG. EARNINGS	O.T. EARNINGS	GROSS PAY	F.I.C.A.	WITHHOLDING	MED. INSURANCE	BONDS	STOCK	PENSION
420.00	141.84	561.84	37.64	72.31	19.93	0	0	16.88

STATE WITHHOLDING		DISABILITY INS.		TOTAL NET PAY
11.25		3.86		399.97

YEAR TO DATE TOTALS

REG. EARNINGS	O.T. EARNINGS	GROSS PAY	F.I.C.A.	WITHHOLDING	MED. INSURANCE	BONDS	STOCK	PENSION
2,100.00	712.50	2,812.50	188.44	361.55	49.65	0	0	84.40

STATE WITHHOLDING		DISABILITY INS.		TOTAL NET PAY TO DATE
56.25		19.30		2,052.91

1. How much money was deducted from this employee's check for the current pay period?

2. How much has this employee paid for medical insurance so far this year?

3. How much money was deducted for the employee's share of the pension plan during this pay period?

4. If the employee had $19.25 deducted for a savings bond, what would the new net pay be?

5. What would the employee's gross pay have been if the overtime had been only $78.80?

6. What is the employee's net pay so far this year?

UNIT 5—PRETEST
MULTIPLYING DECIMALS

Tell how many decimal places each of these numbers has.

1. 47.6133 ....

2. 4 ....

3. 81.6 ....

4. 91.63475 ....

Rewrite and solve these problems.

5. .312 × 7 =

6. 61.728 × .45 =

7. 261.23 × .035 =

8. 6.32 × .727 =

Multiply.

9.   .5
  × .1

10.   .10
  × .9

11.   .35
  × .03

12.   .204
  × .18

Set up the following problems and multiply.

13. 64.12 × .003 =
14. 3.19 × .2806 =
15. 4.28 × 2.8 =
16. .152 × .001 =

Solve the following word problems. Blacken the letter to the right that corresponds to the letter before the correct answer.

17. Bob makes $6.15 an hour cleaning houses. How much does he earn if he works 40 hours?
a. $246   b. $24.60   c. $2,460   d. $2.46

ⓐ ⓑ ⓒ ⓓ

18. Ana lives 1.25 miles from school. How far does she walk to and from school Monday through Friday?
a. 125   b. 12.5   c. 8.6   d. 1.25

ⓐ ⓑ ⓒ ⓓ

MULTIPLYING DECIMALS

Counting Decimal Places

Instruction   Each number to the right of a decimal point is called a **decimal place** or **place**.

Example

> 23.681 has 3 decimal places.
> .04 has 2 decimal places.
> 16 has no decimal places.

Exercise A   Tell how many decimal places each of these numbers has.

1. 26.7 ....

2. 38.33 ....

3. 1.0043 ....

4. 44 ....

5. 26.11467 ....

6. 0.6 ....

7. 77.2 ....

8. 6.28407 ....

9. 3.0349 ....

10. 54 ....

11. 25.736 ....

12. 43.001 ....

13. 19.33 ....

14. 18.211 ....

15. 6 ....

16. 66.21 ....

17. 11.24 ....

18. 3.83 ....

19. 18.345 ....

20. 1.651 ....

21. 46.762 ....

22. 6.2304 ....

23. 74.4146 ....

24. 76.2441 ....

Exercise B   Tell how many decimal places there are in each of these numbers.

1. 18.138 ....

2. 1.2 ....

3. 3.0061 ....

4. 153.26 ....

5. 20 ....

6. 75.0301 ....

7. 101.01 ....

8. 2.7 ....

9. 37.2 ....

10. .23 ....

11. 41.03 ....

12. 8.38 ....

13. 1.96 ....

14. 66 ....

15. 601.4 ....

Courtesy B. F. Goodrich

LESSON TWO:

Placing the Decimal Point in the Product

Instruction

When multiplying decimals, count the number of places in **both** the multiplier and in the multiplicand. Then mark off that many places in the answer, beginning at the right.

Example

```
    .4 1 2      (3 places)
  × .3 5        (2 places)
  2 0 6 0
  1 2 3 6
  .1 4 4 2 0    (5 places)
```

When multiplying decimals, the decimal points do not have to be placed in a straight line.

Example

```
2,316.2 × .053 ──────▶  2,3 1 6.2     (1 place)
                       ×  .0 5 3       (3 places)
                        6 9 4 8 6
                      1 1 5 8 1 0
                      1 2 2.7 5 8 6    (4 places)
```

Exercise A   Rewrite and solve each problem.

1. 2.34 × 7.3 =   2. 32.2 × 47.04 =   3. .736 × 4.2 =

4. 93 × 9.3 =   5. .022 × 9 =   6. .384 × 2.1 =

Exercise B   Rewrite each problem and multiply.

1. 3.26 × 6.4 =   2. .327 × .81 =   3. 42.006 × .21 =

4. 5.096 × .9 =   5. .017 × .17 =   6. 60.024 × 483 =

41

LESSON THREE:

Prefixing Zeros in the Product

Instruction

When multiplying decimals, sometimes there are more decimal places in the multiplier and multiplicand than there are in the product. When this happens, place as many zeros as necessary **in front** of the answer.

Example

```
    .4 1      (2 places)
  × .1 1      (2 places)
    4 1
    4 1
  .0 4 5 1    (4 places)
```

```
    .0 0 4    (3 places)
  ×   .0 3    (2 places)
  .0 0 0 1 2  (5 places)
```

Exercise A   Multiply.

1.   .4
    × .2

2.   .1 6
    × .4

3.   .2 4
    × .0 4

4.   .2 5
    × .3 1

5.   .9
    × .1

6.   .0 0 6
    × .1 3

7.   .4 0 7
    × .1 7

8.   .1 6 4 3
    × .0 0 3

9.   .1 3 5
    × .4 6

Exercise B   Multiply.

1.   .3
    × .3

2.   .1 1
    × .9

3.   .4 5
    × .0 3

4.   .2 6
    × .3 2

5.   .3
    × .2

6.   .4 2
    × .1 1

LESSON FOUR: Solving Word Problems Involving Multiplication of Decimals

Directions   Solve the following word problems.

1. Clyde makes $4.30 an hour at the paper mill. How much does he make in an 8-hour workday?   _____

2. How much does Clyde make a week if he works 40 hours?   _____

3. Alberta buys 3 gallons of milk each week. What is her milk bill each week if milk costs $2.75 a gallon?   _____

4. Laurie Milne's car insurance costs $32.50 per month. What does she pay annually for car insurance?   _____

5. Calvin gives his 5 children their lunch money each Monday morning. It costs $5.50 a week for each child to eat lunch at school. How much lunch money does Calvin need each Monday?   _____

6. Judge Fortner lives 8.3 miles from the courthouse. How many miles does she drive going to and from the courthouse Monday through Friday?   _____

7. Ann Layton's monthly salary is $860.56. What is her annual salary if she is paid every month of the year?   _____

8. A factory worker can buy his lunch from the catering service in the factory for $2.75. How much would he pay for lunch in October if he works 22 days?   _____

9. If you are mailing checks to 8 creditors, how much money do you need to buy one $.20 stamp for each envelope?   _____

10. Silver City has a $.02 sales tax on every dollar's worth of furniture bought in the city. How much sales tax would be charged on furniture costing $357.15?   _____

UNIT 5—REVIEW

Exercise A   Multiply.

1.   .1 2 5
   × .4

2.   1 2.1 0
     × 5

3.   4.8 2
   × 3.0 6

4.   3.1 5
   × 2.8

5.   .1 6 3
   × .2 1

6.   1.0 0 4
   × 8.9

7.   .0 0 3
  × .1 1 2

8.   6.0 3 4
    × 2

9.   .3 4 6
   × .0 1

10.   1 6
   × .2 1

11.   2 4.6 1
  × .0 0 3

12.   .1 0 0 6
   × 8.2

13.   1 6.4
  × 3.5

14.   2 4.1
  × 1.3 4

15. 1 8.2
  × 6

16.   2 0.4
  × 2.4

17. .0 0 6
  × .4

18. 1.2 8
  × 9

Exercise B   Multiply.

1.   1.6 1 4
    × 2 0

2.   .4 3 8
    × .3 3

3.   3.0 0 6
    × .1 5

4.   2.1 6
    × 3 6

5.   .2
    × .3

6.   .4 3
    × .1 2

7.   1 8.2 1
    × 2

8.   2 3
    × .0 1

9.   .6 3
    × .5

10.   9.3
    × 2.6

11.   1 7.9 3
    × .3

12.   .2 0
    × .1 0 1

13.   1.4 0 3
    × .1

14.   1 0 0
    × .6

15.   2.4
    × 2 4

16.   4 2 5.1
    × .9 6

17.   8.4 1 3
    × .2 0 2

18.   7 6 5.4
    × 9.7 6

INSURANCE PREMIUMS

Insurance coverage should be a planned item in your budget. Many different types of policies are available. They can be paid in various ways—annually (once a year), semi-annually (twice a year), quarterly (4 times a year), or as you and your agent agree.

Phillip McLemore chooses to pay for his insurance in the following way.

hospitalization—$27.50 monthly
auto insurance—$96.25 quarterly
life insurance—$134.50 semi-annually

1. What does he pay annually for hospitalization insurance?

----

2. What does he pay annually for auto insurance?

----

3. What does he pay annually for life insurance?

----

4. What does he pay altogether for insurance each year?

----

5. Mr. McLemore has an annual net income of $19,680. What fractional part of his net income goes for insurance annually?

Other items in your budget can also be planned. For example, some expenses come infrequently, such as annually or semi-annually. For these payments, you may want to set aside enough each month to be prepared for these items.

Phillip must pay property tax of $976.08 annually and homeowner insurance of $239.10 semi-annually.

6. How much should be saved each month to pay the property tax?

----

7. How much should be saved each month to pay the homeowner insurance?

46

UNIT 6—PRETEST
DIVIDING DECIMALS

Work the following problems.

1. $2\overline{)8.8}$

2. $22\overline{)46.2}$

3. $30\overline{)66.24}$

4. $14\overline{)4.62}$

Divide.

5. $2\overline{)1}$

6. $5\overline{)1.8}$

7. $16\overline{)6.88}$

8. $6\overline{)5.19}$

Divide the following decimals.

9. $1.2\overline{).36}$

10. $.16\overline{).96}$

11. $11.7\overline{)9.36}$

12. $.21\overline{)42.651}$

Divide.

13. $.9\overline{)909}$

14. $3.2\overline{)416}$

15. $.0015\overline{)123}$

16. $1.1\overline{)1,210}$

Divide. Round quotients off to the nearest hundredth.

17. $.62\overline{)4.468}$ 18. $8\overline{)60}$

19. $9.7\overline{)1,411}$ 20. $120\overline{)20.9}$

Read and solve each problem. Blacken the letter to the right that corresponds to the letter before the correct answer.

21. Alton paid $17.37 a month for a sewing machine. The total cost was $416.88. How many payments did he make?
    a. 24   b. 48   c. 36   d. 28

    [a] [b] [c] [d]

22. Sandra has budgeted $110 for entertainment for four weeks. How much can she spend each week if she spends an equal amount during each 7-day period?
    a. $17.50   b. $2.75   c. $27.50   d. $275

    [a] [b] [c] [d]

23. Daniel has chosen to join a professional organization whose dues are $175 annually. The dues will be deducted from his pay in seven equal payments. What will be the amount of each payment?
    a. $26   b. $25.50   c. $27   d. $25

    [a] [b] [c] [d]

24. A discount store sells toddler-sized disposable diapers in a box of twelve for $2.99. What is the cost per diaper, rounded off to the nearest cent?
    a. 27¢   b. 30¢   c. 25¢   d. 24¢

    [a] [b] [c] [d]

Divide to compute the unit prices of the following items. Circle the item in each pair that is the better buy.

25. a. Sam's Sausage: 16 oz. for $1.49

    b. Sam's Sausage: 24 oz. for $2.19

26. a. Howard D. Bacon: 12 oz. for $1.79

    b. Parker Bacon: 16 oz. for $2.55

27. a. Sparkle Cola: 12 oz. for $.39

    b. Cooler Cola: 32 oz. for $.75

28. a. Kleen Detergent: 70 oz. for $7.67

    b. Kleen Detergent: 34.5 oz. for $3.35

DIVIDING DECIMALS

LESSON ONE:
Dividing a Decimal by a Whole Number

Instruction

When dividing, if there is no decimal in the divisor, (1) move the decimal in the dividend straight up into the quotient, and (2) work the problem.

Example

$$5\overline{)2.5} \longrightarrow 5\overline{)2.5} \longrightarrow 5\overline{)2.5} \atop -25 \atop {}^{.5}$$

NOTE: To the right of the decimal point in the quotient, there must be a number above each number in the dividend. Zeros must be written in the quotient when no other numbers are.

$$15\overline{)1.05} \longrightarrow 15\overline{)1.05} \longrightarrow 15\overline{)1.05} \atop -105 \atop {}^{.07}$$

Exercise A

Work the following problems. Be sure to put the decimal point in the answer before dividing.

1. $3\overline{)6.6}$ 2. $16\overline{)1.44}$ 3. $22\overline{)24.2}$

4. $18\overline{)5.4}$ 5. $31\overline{).217}$ 6. $24\overline{)7.68}$

Exercise B

Divide.

1. $4\overline{)16.4}$ 2. $44\overline{)9.24}$ 3. $16\overline{)502.4}$

49

LESSON TWO:
Dividing a Small Number by a Larger Number

Instruction

Every whole number has a decimal point, but usually you never see it. This decimal point is useful in division when the dividend is smaller than the divisor

In this type of division, it is necessary to put zeros in the dividend. To solve this type of division, follow these steps.

(1) Place the decimal point to the right of the last digit in the dividend.

(2) Move the decimal point straight up into the quotient.

(3) Work the problem.

(4) Annex as many zeros as necessary in the dividend and continue dividing until you get a remainder of zero.

Example

$$25 \overline{)5} \qquad 25 \overline{)5.} \qquad \begin{array}{r} .2 \\ 25 \overline{)5.0} \\ -50 \end{array}$$

$$8 \overline{)3} \qquad 8 \overline{)3.} \qquad \begin{array}{r} .375 \\ 8 \overline{)3.000} \\ -24 \\ \hline 60 \\ -56 \\ \hline 40 \\ -40 \end{array}$$

Exercise A   Divide.

1. $4 \overline{)1}$

2. $5 \overline{)9}$

3. $5 \overline{)6}$

4. $17 \overline{)6.8}$

5. $8 \overline{)7}$

6. $14 \overline{)3.85}$

7. $26 \overline{)5.2}$

8. $12 \overline{)6}$

9. $15 \overline{).045}$

10. $16\overline{)2.448}$ 11. $10\overline{)9.9990}$ 12. $12\overline{)10.008}$

13. $9\overline{)1.305}$ 14. $6\overline{)5.4}$ 15. $3\overline{)2.43}$

16. $5\overline{)3.5325}$ 17. $7\overline{)4.2}$ 18. $15\overline{)4.545}$

Exercise B   Divide.

1. $8\overline{)5}$ 2. $2\overline{)41}$ 3. $16\overline{)4}$

4. $15\overline{)5.4}$ 5. $75\overline{)60}$ 6. $25\overline{)1}$

7. $80\overline{)75.20}$ 8. $9\overline{)4.5}$ 9. $47\overline{)23.5}$

LESSON FOUR: Dividing a Decimal by a Decimal

Instruction

When dividing, if there is a decimal point in the divisor, follow these steps.

(1) Get rid of the decimal point in the divisor by moving it to the right as far as possible and marking its location with a caret (∧).

(2) Move the decimal in the dividend to the right as many places as you moved it in the divisor and mark its location with a caret.

(3) Move the decimal point straight up into the quotient.

(4) Work the problem.

Example

$$.25\overline{)\,.225} \longrightarrow .25_\wedge\overline{)\,.22_\wedge5} \longrightarrow .25_\wedge\overline{)\,.22_\wedge5} \quad \begin{array}{r}.9\\-225\end{array}$$

Exercise A   Divide.

1. $.6\overline{)\,1.8}$

2. $.12\overline{)\,.48}$

3. $.08\overline{)\,.7832}$

4. $1.3\overline{)\,.39}$

5. $6.9\overline{)\,35.88}$

6. $1.28\overline{)\,.1536}$

Exercise B   Divide.

1. $.3\overline{)\,9.9}$

2. $.14\overline{)\,.42}$

3. $.23\overline{)\,84.87}$

4. $14.3\overline{)\,5.72}$

YOU PAY ONLY
$1.86

52

Dividing a Whole Number by a Decimal

Instruction

If you have a decimal point in the divisor but not in the dividend, follow these steps.

(1) Place a decimal point to the right of the last number in the dividend.

(2) Get rid of the decimal point in the divisor by moving it to the right.

(3) Move the decimal point in the dividend to the right the same number of places by annexing as many zeros as necessary.

(4) Move the decimal point straight up into the quotient.

(5) Work the problem.

Example

$$
.25\overline{)5} \longrightarrow .2\,5_\wedge\overline{)5.0\,0_\wedge} \longrightarrow
\begin{array}{r}
2\,0. \\
.2\,5_\wedge\overline{)5.0\,0_\wedge} \\
-5\,0 \\
\hline
0 \\
-\,0 \\
\hline
\end{array}
$$

$$
3.2\overline{)9\,6} \longrightarrow 3.2_\wedge\overline{)9\,6.0_\wedge} \longrightarrow
\begin{array}{r}
3\,0. \\
3.2_\wedge\overline{)9\,6.0_\wedge} \\
-9\,6 \\
\hline
0 \\
-\,0 \\
\hline
\end{array}
$$

Exercise A   Divide.

1. $.18\overline{)54}$

2. $2.6\overline{)52}$

3. $.025\overline{)5}$

4. $.4\overline{)14}$

5. $.013\overline{)39}$

6. $.134\overline{)67}$

Exercise B   Divide.

1. $.9\overline{)27}$

2. $.36\overline{)108}$

3. $.009\overline{)36}$

**Annexing Zeros in Dividends as Necessary
to Complete a Division and Rounding Off
Nonterminating Decimal Quotients**

Instruction

When you get a remainder when dividing, it is sometimes more desirable to annex zeros to the dividend and continue the division rather than leave the remainder. You can do this to carry the division process out to as many places as necessary to get a remainder of zero.

Example

$$
\begin{array}{r} 5 \\ 8\overline{)46} \end{array}
\qquad
\begin{array}{r} 5 \\ 8\overline{)46} \\ -40 \\ \hline 6 \end{array}
\qquad
\begin{array}{r} 5.75 \\ 8\overline{)46.00} \\ -40 \\ \hline 60 \\ -56 \\ \hline 40 \\ -40 \end{array}
\qquad
\begin{array}{r} \\ 2.4\overline{)9} \end{array}
\qquad
\begin{array}{r} 3. \\ 2.4\overline{)9.0} \\ -72 \\ \hline 18 \end{array}
\qquad
\begin{array}{r} 3.75 \\ 2.4\overline{)9.000} \\ -72 \\ \hline 180 \\ -168 \\ \hline 120 \\ -120 \end{array}
$$

Many times, you will never get a remainder of zero, no matter how far you carry out the division. Quotients that do not **terminate** (have a remainder of zero) may be rounded off, usually to the nearest tenth, hundredth, or thousandth. To round off a decimal, follow these steps.

(1) Carry out the division to one more place than you need in your answer.

(2) If the digit that is to be dropped is less than 5, drop it and make no other change.

(3) If the digit to be dropped is 5 or more, drop it and increase the digit to its left by 1.

Example

$$
\begin{array}{r} .4285 \\ 7\overline{)3.0000} \\ -28 \\ \hline 20 \\ -14 \\ \hline 60 \\ -56 \\ \hline 40 \\ -35 \\ \hline 5 \end{array}
$$

.4285 rounded to the nearest tenth = .4

.4285 rounded to the nearest hundredth = .43

.4285 rounded to the nearest thousandth = .429

Exercise A

Divide. Round quotients off to the nearest hundredth.

1. $7\overline{)4}$

2. $2.6\overline{).812}$

3. $9.9\overline{)181}$

4. $330\overline{)107}$

5. $8.7\overline{)281.5}$

6. $12\overline{)49}$

Exercise B

Divide. Round off quotients to the nearest tenth.

1. $.22\overline{)4.46}$

2. $12\overline{)50}$

3. $13\overline{)16}$

4. $.67\overline{)8.962}$

5. $48\overline{)806.3}$

6. $61\overline{)4.921}$

7. $97\overline{)107.32}$

8. $220\overline{)220.2}$

9. $6\overline{)4.372}$

Solving Word Problems Involving Division of Decimals

Instruction

Figuring out how to solve word problems involving division is sometimes confusing. Here are some word problems and suggested ways of thinking about them.

Example

Carlene buys four loaves of bread for $2.12. How much did each loaf cost?

THINK: The total amount would be divided equally by four items.

THEREFORE: $2.12 would be divided by 4.

SOLUTION:

$$\begin{array}{r} \$\ .5\,3 \\ 4\overline{)\,\$\,2.1\,2} \\ -2\,0 \\ \hline 1\,2 \\ -1\,2 \\ \hline \end{array}$$

Harry knows that he has to pay a loan company $16.50 a month until he has paid $132. How many monthly payments of $16.50 does he have to make?

THINK: He must pay $16.50 enough times until the total paid is $132.

THEREFORE: He needs to find how many 16.50s there are in 132.

SOLUTION:

$$\begin{array}{r} 8. \\ 16.5\,0_\wedge\overline{)\,1\,3\,2.0\,0_\wedge} \\ -1\,3\,2\,0\,0 \\ \hline \end{array}$$

Directions

Read and solve each problem.

1. If you take a job paying $12,606.60 a year, what will your monthly salary be? --------------------

2. You take your husband and two children out for lunch. The total bill is $13. If everyone ordered the same thing, what was the cost of the meal for each person? --------------------

3. Mickey's car holds 21.5 gallons of gasoline. If he can drive 301 miles on that amount, how many miles per gallon does his car get? --------------------

4. A grocery store advertises a case of dog food (24 cans) for $6.96. How much does each can cost? ---------------------

5. Mrs. Dunn pays $25.50 each month for life insurance. For how many months is she paid up if she has paid $127.50? ---------------------

6. Carla's annual union dues are $189. If she pays the same amount each month, what will her monthly union dues be? ---------------------

7. Mr. Swift bought a 13.5 pound turkey for $12.83. What was the price per pound? ---------------------

8. The gas station attendant fills your car with 18.5 gallons of gasoline and charges you $18.13. What is the price per gallon? ---------------------

9. Milton figures that it takes 1.4 yards of fabric to cover a sofa pillow. How many pillows can he cover with 5.6 yards of material? ---------------------

10. A mechanic makes $62.80 for working 8 hours. How much does she make each hour? ---------------------

11. Three cans of green beans cost $1.59. What is the price per can? ---------------------

12. Jill and Joe rented a canoe for 4 hours for $35. What was the rental charge per hour? ---------------------

LESSON SEVEN:

Solving Word Problems Involving Multiplication and Division of Decimals

Directions

Read and solve each problem.

1. Ted Gleason read in the newspaper that during the past year consumer prices had increased 12.4 percent. If last year he paid $575.00 for a television, what could he expect to pay this year?

   ....

2. Teresa Oliveira is a bookkeeper. She needs to prepare the sales tax report for January. There were two taxable sales during the month. The first sale made was in the city, where the tax is $.055 for every dollar sold. The amount of the sale was $2,450.00. The second sale, amounting to $1,575.00 was made outside the city, where the tax is $.04. How much tax was collected to be deposited with the report?

   ....

3. Susan Hellums is getting ready to send out the mail. She has 18 letters needing $.20 postage each and 9 letters needing $.37 postage each. How much is the total postage?

   ....

4. The price of a 3-pound can of coffee is $7.98. Bulk coffee is available in 5-pound sacks for $13.45. Which of the above packages is the better buy?

   ....

5. John Hunt's car gets 19.1 miles per gallon of gas. How far can he drive on 500 gallons of gas?

   ....

UNIT 6—REVIEW

When dividing decimals, it is best to first look at the problem and decide what kind of situation you have. Then work the problem.

SITUATION: no decimal in the divisor

(1) Find the decimal point in the dividend.
(2) Move the decimal point straight up into the quotient.
(3) Work the problem.

Example

$$25\overline{)5} \longrightarrow 25\overline{)5.} \longrightarrow 25\overline{)5.0} \quad \begin{array}{r} .2 \\ -50 \end{array}$$

$$25\overline{).5} \longrightarrow 25\overline{).50} \longrightarrow 25\overline{).50} \quad \begin{array}{r} .02 \\ -50 \end{array}$$

$$25\overline{)5.1} \longrightarrow 25\overline{)5.1} \longrightarrow 25\overline{)5.100} \quad \begin{array}{r} .204 \\ -50 \\ \hline 100 \\ -100 \end{array}$$

SITUATION: decimal in the divisor

(1) Get rid of the decimal point in the divisor by moving it to the right.
(2) Move the decimal point an equal number of places to the right in the dividend.
(3) Move the decimal straight up into the quotient.
(4) Work the problem.

Example

$$.25\overline{).25} \qquad .25_\wedge\overline{).25_\wedge} \qquad .25_\wedge\overline{).25_\wedge} \quad \begin{array}{r} 1. \\ -25 \end{array}$$

$$.25\overline{).5} \qquad .25_\wedge\overline{).50_\wedge} \qquad .25_\wedge\overline{).50_\wedge} \quad \begin{array}{r} 2. \\ -50 \end{array}$$

$$.25\overline{)5} \qquad .25_\wedge\overline{)5.00_\wedge} \qquad .25_\wedge\overline{)5.00_\wedge} \quad \begin{array}{r} 20. \\ -50 \end{array}$$

$$\overset{}{.25\overline{)250}} \longrightarrow \overset{}{.25_{\wedge}\overline{)250.00_{\wedge}}} \longrightarrow \begin{array}{r} 1000. \\ .25_{\wedge}\overline{)250.00_{\wedge}} \\ -25 \\ \hline 0 \\ -0 \\ \hline 0 \\ -0 \\ \hline 0 \\ -0 \\ \hline \end{array}$$

Exercise A   Divide.

1. $2\overline{)6.4}$

2. $3\overline{).84}$

3. $14\overline{)7}$

4. $15\overline{)6}$

5. $12\overline{)14.4}$

6. $2.4\overline{)50.4}$

7. $.15\overline{)45}$

8. $3.2\overline{).96}$

9. $.15\overline{).225}$

Exercise B   Divide.

1. $3\overline{)3.9}$

2. $5\overline{).65}$

3. $12\overline{)6}$

4. $16\overline{)4}$

5. $22\overline{).44}$

6. $2.55\overline{)10.2}$

Exercise C   Divide.

1. $1.8\overline{)\,.4\,3\,2}$

2. $12\overline{)\,1\,9.2}$

3. $.1\,3\,4\overline{)\,.6\,7}$

4. $.2\,2\overline{)\,4\,4}$

5. $.0\,0\,3\overline{)\,1\,1\,9.1}$

6. $2\,4\overline{)\,1\,8}$

7. $.1\,5\overline{)\,3}$

8. $2\overline{)\,.0\,7\,2}$

9. $.0\,3\overline{)\,9}$

10. $4.9\overline{)\,1.2\,2\,5}$

11. $.2\,6\overline{)\,2\,6\,2\,6}$

12. $.3\,1\overline{)\,1\,0\,0.4\,4}$

UNIT PRICING

How many times have you stood in a grocery store attempting to select the "best buy" from the shelf? Sometimes you prefer a special brand or a certain size, but often you want the least expensive product.

Unit pricing is one of the best savings tools a shopper has. A product's unit price is how much the item costs per pound, per ounce, per pint, etc. Since the same product may come in packages of different sizes, the only way to know which costs less is to compare the unit prices.

You can figure unit prices by dividing the price of the item by the amount (contents) in the package. Some stores display unit prices for their products so you do not have to do the figuring. Pocket calculators can make the figuring easier to do.

Example

A 12-ounce can of orange juice sells for 99¢. A 6-ounce can sells for 54¢. Which is the better buy?

$$\text{unit price} = \text{price} \div \text{contents}$$

large can

$$\begin{array}{r} \$.0825 = \$.083 = 8.3¢ \\ 12\overline{)\$.990} \\ \underline{-0} \\ 99 \\ \underline{-96} \\ 30 \\ \underline{-24} \\ 60 \\ \underline{-60} \\ 0 \end{array}$$

small can

$$\begin{array}{r} \$.09 = 9¢ \\ 6\overline{)\$.54} \\ \underline{-0} \\ 54 \\ \underline{-54} \\ 0 \end{array}$$

8.3¢ per ounce is less than 9¢ per ounce. The larger can is a better buy.

When figuring unit prices, it is important to be sure that the same units are being used for each item. For example, when comparing the price of a 12-ounce jar of peanut butter and a 3-pound jar, you need to convert 3 pounds to 48 ounces and find the cost **per ounce** for each jar.

Directions

Compute the unit prices of the following items to determine the better buy (based on unit price) for each product. The unit price should be rounded off to the nearest one-tenth cent. Circle the item in each pair that is the better buy.

	Item	Price	Quantity	Unit Price
	a. Tastee Coffee	$2.29	16 oz.	
1.	b. Tastee Coffee	$6.69	3 lb.	
	a. Nutty Peanut Butter	$1.89	18 oz.	
2.	b. Nutty Peanut Butter	$2.89	28 oz.	
	a. Crispy Cereal	$1.77	18 oz.	
3.	b. Crispy Cereal	$.89	8 oz.	
	a. Spicy Catsup	$.57	14 oz.	
4.	b. Spicy Catsup	$.98	32 oz.	
	a. Picnic Ham	$6.97	5.83 lb.	
5.	b. Boneless Ham	$13.60	4.55 lb	
	a. Bacon	$1.59	12 oz.	
6.	b. Bacon	$4.37	2 lbs.	
	a. Laundry Detergent	$3.99	30 oz.	
7.	b. Laundry Detergent	$7.79	70 oz.	
	a. Chip Dip	$.49	3 oz.	
8.	b. Chip Dip	$.77	10 oz.	
	a. Dill Pickles	$1.59	24 oz.	
9.	b. Dill Pickles	$2.46	40 oz.	
	a. Cooking Oil	$1.59	24 oz.	
10.	b. Cooking Oil	$2.84	48 oz.	

UNIT 7—PRETEST
FRACTIONS AND DECIMALS

Change each fraction to its decimal equivalent.

1. $\frac{3}{5}$ 2. $\frac{7}{12}$ 3. $\frac{5}{4}$ 4. $\frac{9}{16}$

Change each fraction into its decimal equivalent. Write remainders as fractions.

5. $\frac{1}{7}$ 6. $\frac{1}{9}$ 7. $\frac{5}{6}$ 8. $\frac{2}{3}$

Change each fraction into its decimal equivalent. Write remainders as tractions.

9. $14\frac{1}{4}$ 10. $808\frac{4}{5}$

Solve the following problems. Give each answer in decimal form.

11. $619 + 2\frac{2}{5}$ 12. $22.4 - 3\frac{1}{4}$

13. $72.32 \div \frac{4}{5}$ 14. $4.21 \times 4\frac{2}{5}$

Change each decimal into its fractional equivalent. Reduce, if possible.

15. .34   16. .005   17. 666.02   18. 9.007

Change each decimal into its common fraction equivalent.

19. $.66\frac{2}{3}$ 20. $.04\frac{1}{6}$ 21. $.22\frac{3}{4}$ 22. $.5\frac{2}{5}$

Solve the following word problems. Blacken the letter to the right that corresponds to the letter before the correct answer.

23. Glenda bought $\frac{1}{4}$ pound of candy for $4.59 a pound. How much did the candy cost?
    a. 15¢   b. $1.15   c. $1.05   d. $1.25   [a] [b] [c] [d]

24. Dane bought a piece of carpet $3\frac{5}{6}$ feet wide. His hallway is only 3.75 feet wide. How much must he cut off the width of the carpet?
    a. .0083 feet   b. .83 feet   c. .083 feet   d. 8.3 feet   [a] [b] [c] [d]

FRACTIONS AND DECIMALS

Writing Fractions as Decimals

Instruction

Any fraction can be written as a decimal. Any decimal can be written as a fraction. This is easier to understand when you think in terms of money.

Example

Fraction		Decimal
$\frac{1}{2}$ dollar	=	$.50
$\frac{1}{4}$ dollar	=	$.25

Any fraction can be changed into its decimal equivalent by dividing the numerator by the denominator. Remember, zeros may be placed to the right of the decimal point as needed in order to carry out the division process.

Example

$$\frac{1}{2} \longrightarrow 2\overline{)1.0} \longrightarrow \frac{1}{2} = .5$$
$$\underline{-10}$$

$$\frac{3}{4} \longrightarrow 4\overline{)3.00} \longrightarrow \frac{3}{4} = .75$$
$$\underline{-28}$$
$$20$$
$$\underline{-20}$$

$$\frac{5}{8} \longrightarrow 8\overline{)5.000} \longrightarrow \frac{5}{8} = .625$$
$$\underline{-48}$$
$$20$$
$$\underline{-16}$$
$$40$$
$$\underline{-40}$$

$$\frac{3}{2} \longrightarrow 2\overline{)3.0} \longrightarrow \frac{3}{2} = 1.5$$
$$\underline{-2}$$
$$10$$
$$\underline{-10}$$

Exercise A

Change each fraction into its decimal equivalent.

1. $\frac{2}{5}$

2. $\frac{1}{16}$

3. $\frac{7}{10}$

4. $\frac{5}{4}$

5. $\frac{7}{8}$

6. $\frac{15}{16}$

7. $\frac{3}{4}$

8. $\frac{7}{16}$

Exercise B

Change each fraction into its decimal equivalent.

1. $\frac{4}{5}$

2. $\frac{3}{16}$

3. $\frac{9}{10}$

4. $\frac{5}{2}$

5. $\frac{5}{8}$

6. $\frac{13}{16}$

7. $\frac{11}{10}$

8. $\frac{7}{5}$

Changing a Fraction into a Decimal When There is a Remainder

Instruction

Any fraction can be changed into a decimal, but sometimes the answer does not work out evenly. When this happens, you get a remainder. When changing a fraction into a decimal, if there are 3 **decimal** places in the quotient and you have a remainder other than zero, follow these steps.

(1) Place the remainder over the divisor.

(2) Reduce the fraction if possible.

(3) Take the first two places in the quotient and put the fraction with it.

Example

$$
\frac{1}{12} \longrightarrow
\begin{array}{r}
.083 = .08\frac{1}{3} \\
12\overline{)1.000} \\
-96 \\
\hline
40 \\
-36 \\
\hline
4 = \frac{4}{12} = \frac{1}{3}
\end{array}
$$

In lesson 5 of chapter 6, you saw another way to handle quotients that do not work out evenly—rounding off. You should be familiar with both processes.

Exercise A

Change each fraction into its decimal equivalent. Write remainders as fractions.

1. $\frac{1}{3}$

2. $\frac{2}{3}$

3. $\frac{1}{6}$

4. $\frac{5}{18}$

5. $\frac{1}{9}$

6. $\frac{5}{6}$

7. $\frac{1}{7}$

8. $\frac{3}{8}$

9. $\frac{2}{7}$

10. $\frac{2}{9}$

Exercise B

Change each fraction into its decimal equivalent. Write remainders as fractions.

1. $\frac{7}{18}$

2. $\frac{5}{12}$

3. $\frac{5}{6}$

4. $\frac{7}{12}$

5. $\frac{11}{19}$

6. $\frac{5}{7}$

7. $\frac{6}{7}$

8. $\frac{8}{9}$

9. $\frac{7}{8}$

10. $\frac{10}{11}$

Changing a Mixed Number into a Mixed Decimal

Instruction

A mixed number can be changed into a mixed decimal. Write the whole number to the left of the decimal point. Convert the fraction into a decimal by dividing the numerator by the denominator. Write the result to the right of the decimal point.

Example

$$23\frac{1}{4} \longrightarrow 4\overline{)\begin{array}{c} .25 \\ 1.00 \\ \underline{-8} \\ 20 \\ \underline{-20} \end{array}} \longrightarrow 23\frac{1}{4} = 23.25$$

Remember, when changing a mixed fraction into a mixed decimal, be sure that the whole number is on the left-hand side of the decimal point.

Exercise A

Change each mixed number into its mixed decimal equivalent. Write remainders as fractions.

1. $5\frac{3}{4}$

2. $25\frac{5}{8}$

3. $13\frac{1}{3}$

4. $3\frac{1}{9}$

5. $148\frac{3}{5}$

6. $17\frac{1}{4}$

7. $10\frac{3}{8}$

8. $7\frac{7}{15}$

9. $20\frac{7}{10}$ **10.** $64\frac{7}{9}$

Exercise B   Change each mixed number into its mixed decimal equivalent.
Write remainders as fractions.

1. $7\frac{1}{5}$ **2.** $14\frac{3}{8}$

3. $52\frac{2}{3}$ **4.** $7\frac{5}{6}$

5. $216\frac{3}{4}$ **6.** $101\frac{1}{8}$

7. $61\frac{2}{3}$ **8.** $17\frac{2}{5}$

9. $48\frac{1}{3}$ **10.** $87\frac{1}{6}$

LESSON FOUR: Solving Problems Containing Both Decimals and Fractions

Instruction

When working with fractions and decimals, sometimes it is easier to solve the problem by changing the fraction into a decimal.

Example

$.43 + \frac{1}{2} \longrightarrow \frac{1}{2} = .5 \longrightarrow$

$$
\begin{array}{r}
.43 \\
+ .5 \\
\hline
.93
\end{array}
$$

$22\frac{3}{4} \times 1.16 \longrightarrow 22\frac{3}{4} = 22.75 \longrightarrow$

$$
\begin{array}{r}
22.75 \\
\times 1.16 \\
\hline
13650 \\
2275 \\
2275 \\
\hline
26.3900
\end{array}
$$

Exercise A

Solve the following problems. Give each answer in decimal form.

1. $18.3 + 14\frac{1}{5}$

2. $3.64 - \frac{1}{20}$

3. $2.12 \times 5\frac{1}{4}$

4. $6.4 \div \frac{4}{5}$

5. $5.16 \times 2\frac{1}{5}$

6. $3.2 \div \frac{1}{2}$

7. $7.46 + 2\frac{4}{5}$

8. $27.5 \times 2\frac{1}{4}$

9. $86.19 - \frac{33}{100}$

10. $31.2 + \frac{3}{4}$

11. $11.6 \times 8\frac{4}{5}$ **12.** $9.9 \div \frac{1}{20}$

Exercise B
Solve these problems. Give each answer in decimal form.

1. $29.9 + 4\frac{3}{5}$ **2.** $8.16 - \frac{1}{25}$

3. $21\frac{2}{5} \times 4.3$ **4.** $15\frac{1}{8} \div 5$

5. $4.51 + 16\frac{1}{5}$ **6.** $\frac{1}{2} - .06$

7. $3.08 \div \frac{1}{4}$ **8.** $2.1414 \times \frac{2}{5}$

9. $47.1 \div \frac{3}{4}$ **10.** $\frac{3}{4} - .673$

11. $\frac{1}{4} - .18$ **12.** $6.4 + \frac{1}{2}$

Changing Decimals Into Fractions

Instruction

Every decimal can be written as a fraction. To write a decimal as a fraction, follow these steps.

(1) Read the decimal.
(2) Put the digits to the right of the decimal point in the numerator.
(3) Put 10, 100, 1,000, etc., in the denominator.
(4) Reduce if possible.

Remember, a decimal with one digit has a denominator of 10, a decimal with two digits has a denominator of 100, a decimal with three digits has a denominator of 1,000, and a decimal with four digits has a denominator of 10,000.

Example

.8 $\quad$ eight-tenths $\longrightarrow \frac{8}{10} = \frac{4}{5}$

.04 $\longrightarrow$ four-hundredths $\longrightarrow \frac{4}{100} = \frac{1}{25}$

.625 $\longrightarrow$ six hundred twenty-five thousandths $\longrightarrow \frac{625}{1000} = \frac{5}{8}$

4.05 $\longrightarrow$ four and five-hundredths $\longrightarrow 4\frac{5}{100} = 4\frac{1}{20}$

Exercise A

Change each decimal into its fractional equivalent. Reduce if possible.

1. .1
2. .38
3. .4
4. .025
5. .25
6. 338.04
7. .15
8. .05
9. .6
10. .75
11. .32
12. .03
13. .45
14. .2

Exercise B

Change each decimal into its fractional equivalent. Reduce if possible.

1. .7
2. .085
3. .8
4. 547.15
5. .003
6. 6.625
7. 64.1
8. .0025
9. 8.05
10. .01
11. 45.5
12. .005

Changing Decimals into Fractions When the Decimal Contains a Fraction

Instruction

Every decimal can be written as a fraction. Even a decimal with a fraction can be changed into a common fraction. Follow these steps.

(1) Read the decimal.

(2) Put the digits which are to the right of the decimal point in the numerator.

(3) Put either 10, 100, 1,000, etc., in the denominator.

(4) Change the mixed number in the numerator into an improper fraction.

(5) Divide the numerator by the denominator (10, 100, 1,000, etc.).

(6) Reduce if possible.

Example

Write $.18\frac{1}{3}$ as a fraction.

Step 1 $.18\frac{1}{3}$ = eighteen and one-third hundredths

Step 2 $\dfrac{18\frac{1}{3}}{}$

Step 3 $\dfrac{18\frac{1}{3}}{100}$

Step 4 $\dfrac{\frac{55}{3}}{100}$

Step 5 $\frac{55}{3} \div 100 = \frac{55}{3} \times \frac{1}{100} = \frac{55}{300}$

Step 6 $\frac{55}{300} = \frac{11}{60}$

$.18\frac{1}{3} = \frac{11}{60}$

Exercise A

Change each decimal into its common fractional equivalent.

1. $.33\frac{1}{3}$

2. $.66\frac{2}{3}$

3. $.11\frac{1}{9}$

4. $.25\frac{2}{7}$

5. $.20\frac{1}{3}$ **6.** $.08\frac{2}{3}$

7. $.05\frac{1}{4}$ **8.** $.5\frac{1}{8}$

Exercise B Change each fraction into its common fraction equivalent.

1. $.16\frac{2}{3}$ **2.** $.04\frac{1}{6}$

3. $.22\frac{1}{2}$ **4.** $.62\frac{1}{3}$

5. $.6\frac{1}{4}$ **6.** $.8\frac{2}{5}$

LESSON SEVEN:
Solving Word Problems Involving Decimals and Fractions

Instruction

When solving problems involving decimals and fractions, write the decimal as its fractional equivalent, or write the fraction as its decimal equivalent—whichever seems easier.

Directions

Solve the following word problems.

1. Karla gets $8.55 per hour when she works overtime. If she works $3\frac{1}{4}$ hours overtime, how much money will she make?

   ....

2. Mr. Williams determined that he needs $6\frac{1}{2}$ yards of material to make doll dresses. If the material he chooses costs $1.48 a yard, how much will the material cost him?

   ....

3. Charles filled his car with $15\frac{3}{4}$ gallons of gasoline. He owed the attendant $15.44. What was the price of the gasoline per gallon?

   ....

4. A painter mixes 4.33 gallons of paint and uses $1\frac{3}{5}$ gallons in the hall. How much paint does she have left?

   ....

5. Clifford bought $15\frac{1}{2}$ pounds of nuts for $.98 a pound. How much did the nuts cost?

   ....

6. An electrician needs to move an electrical outlet $\frac{4}{5}$ of a foot. She has .366 foot of wire. How much more wire does she need?

   ....

7. A machinist has to drill a hole through 2.6 inches of steel. He has drilled $1\frac{1}{4}$ inches. How much farther must he drill?

   ....

76

UNIT 7—REVIEW

Exercise A   Complete the following statements.

1. To change any fraction into its decimal equivalent, divide the _____

   by the _____.

2. To change any decimal into its fractional equivalent, place the digits which are

   to the right of the decimal point in the _____ of the fraction.

3. A decimal with one digit has a denominator of _____.

4. A decimal with two digits has a denominator of _____.

5. A decimal with three digits has a denominator of _____.

6. A decimal with four digits has a denominator of _____.

7. A fraction should be _____ to its lowest term.

8. Every fraction can be written as a _____.

9. Every decimal can be written as a _____.

Exercise B   Complete the following chart.

	Fraction	Decimal
1.	$\frac{1}{4}$	
2.	$4\frac{2}{5}$	
3.	$\frac{1}{6}$	
4.		3.75
5.		$.33\frac{1}{3}$
6.	$\frac{5}{4}$	
7.		.0625
8.		.15
9.	$\frac{1}{40}$	
10.		14.16
11.	$2\frac{1}{5}$	

77

Exercise C Complete the following chart.

	Fraction	Decimal
1.		.75
2.	$18\frac{1}{5}$	
3.	$\frac{1}{3}$	
4.		.875
5.		$.12\frac{2}{3}$
6.	$\frac{1}{6}$	
7.		5.005
8.		.085
9.	$\frac{7}{4}$	
10.	$\frac{5}{8}$	
11.	$3\frac{3}{5}$	
12.		$.66\frac{2}{3}$
13.	$\frac{7}{5}$	
14.		9.85
15.	$\frac{1}{50}$	
16.	$\frac{1}{2}$	
17.		.015
18.		1.006
19.		$2.33\frac{1}{3}$
20.	$\frac{8}{5}$	
21.	$\frac{1}{25}$	
22.		.01
23.	$\frac{7}{4}$	
24.	$\frac{1}{11}$	
25.		9.25
26.	$3\frac{1}{11}$	
27.		.11

OVERTIME PAY

Most companies have a set number of hours for each employee to work each week. Many companies have a 40-hour work week. If an employee works more than 40 hours, he or she may be paid extra—time and a half, double time, or triple time.

Example

Pay Figured on 40-Hour Work Week

HOURS WORKED	REGULAR HOURLY PAY	OVERTIME PAY RATE
45	$3.50	$1\frac{1}{2}$

REGULAR PAY

$$\begin{array}{r} \$3.50 \\ \times\,40 \\ \hline \$140.00 \end{array}$$

OVERTIME PAY

$\$3.50 \times 1\frac{1}{2} = 3.50 \times 1.5 = \$5.25$ per hour

$$\begin{array}{r} \$5.25 \\ \times\,5 \\ \hline \$26.25 \end{array}$$

GROSS PAY

$$\begin{array}{rl} \$140.00 & \text{regular pay} \\ +\;26.25 & \text{overtime pay} \\ \hline \$166.25 \end{array}$$

Directions

Complete the following chart by computing the regular weekly salary based on a 40-hour work week, overtime pay based on the rate given, and the gross pay.

	Regular Hourly Pay	Hours Worked	Overtime Rate	Regular Pay	Overtime Pay	Gross Pay
1.	$3.55	45	$1\frac{1}{2}$			
2.	$3.76	53	$1\frac{1}{2}$			
3.	$4.53	60	2			
4.	$7.60	47	$2\frac{1}{2}$			
5.	$5.25	43	3			
6.	$4.20	48	2			
7.	$3.35	45	$1\frac{1}{2}$			
8.	$5.09	44	2			
9.	$6.75	55	$1\frac{1}{2}$			
10.	$4.87	62	$2\frac{1}{2}$			
11.	$7.38	42	3			

UNIT 8—PRETEST
DECIMALS AND PERCENTS

Write the following using numerals and the percent sign.

1. nineteen and two-thirds percent _____

2. two hundred nine and nine-tenths percent _____

3. four-fifths percent _____

4. twelve percent _____

Write each percent as its decimal equivalent.

5. 19% =   6. 198% =

7. 6.25% =   8. 3.7% =

Change each percent into its decimal equivalent.

9. $6\frac{4}{5}$% =   10. $67\frac{2}{3}$% =

11. $\frac{2}{5}$% =   12. $206\frac{1}{8}$% =

Change each decimal into a percent. When the decimal contains a common fraction, keep it instead of finding its decimal equivalent.

13. .3 =   14. .295 =

15. 1.27 =   16. $.65\frac{1}{4}$ =

Complete the following chart.

	Decimal	Percent
17.	2.7	
18.		4%
19.	$.33\frac{1}{3}$	
20.		250%
21.		$21\frac{3}{5}$%

80

DECIMALS AND PERCENTS

LESSON ONE: **Writing Percents**

Instruction

This sign (%) is read "percent." *Percent* means per hundred. Any percent can be written as a decimal. Any decimal can be written as a percent.

Example

50% is read "fifty percent."

5% is read "five percent."

175% is read "one hundred seventy-five percent."

$5\frac{1}{2}$% is read "five and one-half percent."

$\frac{1}{2}$% is read "one-half percent."

5.4% is read "five and four-tenths percent."

Exercise A   Write each of the following using the percent sign.

1. thirty-four percent _____

2. twelve and one-fourth percent _____

3. three and four-tenths percent _____

4. four-fifths percent _____

5. one hundred sixteen and three-tenths percent _____

Exercise B   Write each of the following using the percent sign.

1. thirteen percent _____

2. six and one-fifth percent _____

3. one and eight-tenths percent _____

4. two-thirds percent _____

5. forty-five and nine-tenths percent _____

6. seventy-eight and one seventh percent _____

81

Changing Percents to Decimals

Instruction

To write a percent as a decimal, follow these steps.

(1) Drop the percent sign.

(2) Move the decimal point two places to the left.

NOTE: You may put in as many zeros as necessary to the left of
the number in order to move the decimal point two places.

Example

75%	⟶	.75.		75% = .75
5%	⟶	.05.		5% = .05
125%	⟶	1.25.		125% = 1.25
10.5%	⟶	.10.5		10.5% = .105
4.6%	⟶	.04.6		4.6% = .046

Exercise A

Write each percent as its decimal equivalent.

1. 23% =   2. 1% =
3. 9.2% =   4. 112% =
5. 10.3% =   6. 36% =
7. 8% =   8. 100% =
9. 8.25% =   10. 9% =
11. 200% =   12. 30% =
13. 33% =   14. 15.5% =
15. 150% =   16. 225% =

Exercise B

Write each percent as its decimal equivalent.

1. 14% =   2. 3% =
3. 4.4% =   4. 147% =
5. 11.5% =   6. 18% =
7. 4% =   8. 200% =
9. 5.75% =   10. 2% =
11. 12.4% =   12. 7.2% =
13. 8% =   14. 62% =
15. 47% =   16. 16.7% =
17. 90% =   18. 82.4% =
19. .2% =   20. 347% =

LESSON THREE: Changing a Percent Which Contains a Fraction into a Decimal

Instruction

Any percent can be written as a decimal. To change a percent which contains a fraction to a decimal, follow these steps.

(1) Change the fraction to its decimal equivalent by dividing the numerator by the denominator.

(2) Drop the percent sign.

(3) Move the decimal point two places to the left.

Example

$8\frac{1}{4}\%$ $\longrightarrow$ 8.25% $\longrightarrow$ .08.25 $8\frac{1}{4}\% = .0825$

$23\frac{1}{2}\%$ $\longrightarrow$ 23.5% $\longrightarrow$ .23.5 $23\frac{1}{2}\% = .235$

$\frac{1}{5}\%$ $\longrightarrow$ .2% $\longrightarrow$ .00.2 $\frac{1}{5}\% = .002$

Remember, some fractions, like $\frac{1}{3}$, $\frac{2}{3}$, $\frac{1}{6}$, $\frac{5}{6}$, etc., cannot be written as pure decimals. You always get a remainder. When converting percents which contain such fractions into decimals, you can use any one of the following procedures.

(1) Write the fraction as a decimal by carrying the division out to three places and rounding the answer off to the nearest hundredth. (See lesson 5 in chapter 6.)

(2) Write the fraction as a decimal by dividing and writing the remainder as a fraction. (See lesson 2 in chapter 7.)

(3) Convert the whole number portion of the percent to a decimal but leave the fractional part written as a fraction.

Example

Write $16\frac{2}{3}\%$ as a decimal.

(1) $16\frac{2}{3}\%$ $\longrightarrow$ 16.67% $\longrightarrow$ .16.67 $16\frac{2}{3}\% = .1667$

(2) $16\frac{2}{3}\%$ $\longrightarrow$ $16.66\frac{2}{3}\%$ $\longrightarrow$ $.16.66\frac{2}{3}$ $16\frac{2}{3}\% = .1666\frac{2}{3}$

(3) $16\frac{2}{3}\%$ $\longrightarrow$ $.16.\frac{2}{3}$ $16\frac{2}{3}\% = .16\frac{2}{3}$

Exercise A

Change each percent into its decimal equivalent.

1. $4\frac{1}{2}\% =$ 2. $3\frac{2}{3}\% =$

3. $56\frac{3}{4}\% =$ 4. $8\frac{4}{5}\% =$

5. $\frac{3}{5}\% =$

6. $112\frac{1}{2}\% =$

7. $\frac{3}{8}\% =$

8. $5\frac{1}{4}\% =$

9. $7\frac{1}{3}\% =$

10. $\frac{1}{2}\% =$

Exercise B

Change each percent into its decimal equivalent.

1. $5\frac{1}{4}\% =$

2. $5\frac{5}{8}\% =$

3. $18\frac{2}{5}\% =$

4. $2\frac{1}{8}\% =$

5. $\frac{3}{4}\% =$

6. $103\frac{1}{2}\% =$

7. $\frac{5}{8}\% =$

8. $4\frac{1}{4}\% =$

9. $8\frac{2}{3}\% =$

SPECIAL SALE

UP TO 50 PERCENT OFF

WIDE SELECTION OF
JEWELRY

ORNAMENTS

AND GIFT ITEMS

Changing Decimals to Percents

Instruction

Any decimal can be written as a percent. To change a decimal to a percent, follow these steps.

(1) Move the decimal point two places to the right.

(2) Place the percent sign to the right of the number. The decimal point usually is not written unless the percent contains a decimal fraction.

NOTE: Put as many zeros as necessary to the right of the number in order to move the decimal point two places.

Example

.25 ⟶ .25.%   .25 = 25%

3.6 ⟶ 3.60.%   3.6 = 360%

$.34\frac{1}{3}$ ⟶ $.34.\frac{1}{3}\%$ $.34\frac{1}{3} = 34\frac{1}{3}\%$

When a number contains a fraction, you can write the fraction as its decimal equivalent before writing the number as a percent.

Example

$.13\frac{1}{4}$ ⟶ .1325 ⟶ .13.25% $.13\frac{1}{4}$ = 13.25%

$.12\frac{1}{2}$ ⟶ .125 ⟶ .12.5% $.12\frac{1}{2}$ = 12.5%

Remember, fractions like $\frac{1}{3}$ can be written as decimals by carrying out the division to three decimal places and rounding the quotient off to the nearest hundredth.

Example

Write $.33\frac{1}{3}$ as a percent.

$$\begin{array}{r} .333 \\ 3\overline{)1.000} \\ -9 \\ \hline 10 \\ -9 \\ \hline 10 \\ -9 \\ \hline 1 \end{array} = .33$$

(1) Convert $\frac{1}{3}$ to its decimal equivalent by carrying out the division to three decimal places and rounding the quotient off to the nearest hundredth.

$.33\frac{1}{3}$ $.33.\frac{1}{3}$ (2) Move the decimal point in $.33\frac{1}{3}$ two places to the right.

$.33\frac{1}{3}$ 33.33%   (3) Replace $\frac{1}{3}$ with its decimal equivalent. Write the percent sign.

85

Exercise A

Change each decimal into a percent. When the decimal contains a common fraction, keep it instead of finding its decimal equivalent.

1. .75 =
2. .1 =
3. .07 =
4. $.16\frac{2}{3}$ =
5. .4 =
6. $.06\frac{1}{4}$ =
7. .325 =
8. 1.36 =
9. $.68\frac{4}{5}$ =
10. .19 =
11. $.48\frac{1}{3}$ =
12. 3.86 =
13. 2.4 =
14. .7 =
15. .09 =
16. $2.33\frac{1}{3}$ =
17. $4.66\frac{2}{3}$ =
18. $.76\frac{1}{4}$ =
19. .214 =
20. .004 =
21. $.22\frac{3}{4}$ =
22. $1.6\frac{1}{2}$ =

Exercise B

Change each decimal into a percent. When the decimal contains a common fraction, keep it instead of finding its decimal equivalent.

1. .28 =
2. .8 =
3. .15 =
4. $.33\frac{1}{3}$ =
5. .9 =
6. $.02\frac{1}{5}$ =
7. .685 =
8. 1.15 =
9. $.25\frac{1}{2}$ =
10. .43 =
11. 2.25 =
12. .714 =
13. .063 =
14. $9.03\frac{1}{2}$ =
15. .47 =
16. .333 =
17. .5 =
18. $.05\frac{1}{4}$ =
19. .85 =

UNIT 8—REVIEW

Exercise A   Complete the following statements by filling in the blanks.

1. This sign (%) is read _____.

2. To change a percent into its decimal equivalent, drop the percent sign and move the decimal point _____ places to the _____.

3. To change a decimal into a percent, move the decimal point _____ places to the _____ and write the percent sign.

Exercise B   Complete the following chart.

	Decimal	Percent
1.		78%
2.	.87	
3.		$12\frac{1}{2}$%
4.		8%
5.	2.5	
6.	$.16\frac{2}{3}$	
7.		625%
8.		$\frac{3}{4}$%
9.	.875	
10.	.3	
11.		24%
12.	1.7	
13.	20.78	
14.		112%
15.	.855	
16.	1.623	
17.		206%

87

Exercise C   Complete the following chart.

	Decimal	Percent
1.		25%
2.	.7	
3.		100%
4.	.92	
5.		3%
6.	$.66\frac{2}{3}$	
7.		$\frac{4}{5}\%$
8.	.375	
9.		$8\frac{1}{4}\%$
10.		$33\frac{1}{3}\%$
11.	.225	
12.	$.81\frac{1}{4}$	
13.		200%
14.	9	
15.		$\frac{3}{4}\%$
16.	2.25	
17.		30%
18.	.33	
19.		$80\frac{1}{2}\%$
20.		$\frac{1}{5}\%$
21.	6.27	
22.	.075	
23.		250%
24.	4.7	
25.		$\frac{1}{4}\%$
26.	.25	
27.	.4	

CREDIT CONTRACTS

There are two basic kinds of credit—**retail credit** and **cash credit**. Retail, or store credit, can be used instead of cash to buy many things. Charge accounts and installment purchases are examples. Cash credit is the actual borrowing of money to purchase items or to pay debts.

Shopping for credit is not easy. You need to be selective in order to find the best deal. Interest rates vary greatly from place to place. Finance charges for cash loans are often lower than finance charges on credit accounts or installment purchases. Finance charges are often lowest at credit unions.

The Truth in Lending Law has made it easier to shop for credit than it used to be. The law requires that you be told the **annual percentage rates** (APR) for loans made to you. For example, a finance charge based on a monthly interest rate of 1.5% on your unpaid balance would mean that the annual percentage rate is 18%. When you are told a percentage rate, always ask if it is the annual percentage rate. The law also requires creditors to tell you the cost of a loan in dollars.

When you borrow money, try to repay it in the fewest number of months possible. You will pay less interest this way, but your monthly payments will be higher. However, don't agree to make high monthly payments if you can't afford them. The chart below shows how the cost of a loan increases as the length of the loan increases.

Number of months of repayment	Annual percentage rate	Finance charge or dollar cost on $350 borrowed	Amount of each monthly payment
12 months (1 year)	12%	$23.20	$31.10
18 months (1½ years)	12%	$34.30	$21.35
24 months (2 years)	12%	$45.28	$16.47
36 months (3 years)	12%	$68.32	$11.62

Before you sign a credit agreement, read it carefully. Never sign a contract with spaces left blank. Be sure you understand everything in the contract. If you don't, have someone explain it to you. Take the contract to a legal service center if you have any doubts.

Look for the following items in credit contracts.

1. **Purchase Price Or Amount Borrowed**

   The purchase price is the actual cost of the item you are buying. The amount borrowed is the actual amount of money being financed.

2. **Annual Percentage Rate (APR)**

   The annual percentage rate must be stated in all credit agreements.

3. **Finance Charges in Dollars**

   The finance charge is the amount you pay in dollars for using credit. You pay this amount in addition to the purchase price or amount borrowed.

4. Cash Down Payment, If Any

You may have to make a down payment when you buy an item or service. You will not have a down payment when you borrow money. The down payment is money you pay toward the purchase of the item or service. It is subtracted from the purchase price. You do not pay interest on the down payment.

5. Trade-in Allowances, If Any

Trade-in allowances apply only when buying items. For example, when buying a car, the dealer may give you a certain amount for your old car. The amount is subtracted from the purchase price of the new car. You pay interest only on the balance. Sometimes, the trade-in allowance may be used as the down payment.

6. Insurance Charges in Dollars, If Any

Some lenders encourage you to take out insurance on the money borrowed. The insurance is usually for life and disability. Credit unions and banks usually do not require you to purchase the insurance. Many finance companies highly encourage it. If you take it, the fee is added to the amount you pay.

7. Service Charges Or Other Costs

A service charge is money you pay to the lender for handling your account. Before you sign a loan agreement, be sure you understand about the service charges. They can sometimes be rather expensive. They add to the cost of the loan. Banks sometimes have service charges for some types of loans. Credit unions usually do not. However, you must be a member of a credit union to borrow money there. Finance companies usually have a service charge.

8. Total Amount Due

The total amount due on any loan includes the following items.
• purchase price minus down payment or trade-in allowance (This is the amount borrowed.)   • finance charges   • insurance charges, if any   • service charges, if any   • any other charges stated in the contract.

The chart below shows the total amount due on a $350 loan for two years. The chart compares charges at a credit union, a bank, and a finance company. Which one offers the best, or cheapest, loan arrangement? _____

	Credit Union APR 12%	Bank APR 11.22%	Finance Company APR 21.52%
Amount of loan	$350.00	$350.00	$350.00
Finance charge on 2-year repayment	45.28	41.84	78.96
Service charge for 2 years	none	8.00	9.12
Life insurance	not required	not required	6.84*
Disability insurance	not required	not required	13.15*
Total cost or amount due	$395.28	$399.84	$458.08

*Not required
but highly encouraged

UNIT 9—PRETEST
DECIMALS AND PERCENTS

Write each fraction as its percent equivalent. If a fraction cannot be changed to a pure decimal (no fractional part), write the percent as a mixed fraction ($33\frac{1}{3}$%).

1. $\frac{7}{10}$ 2. $15\frac{1}{2}$ 3. $\frac{3}{50}$ 4. $17\frac{1}{9}$

Write each percent as its fractional equivalent.

5. 18%   6. 72%   7. 175%   8. 210%

Write these percents as fractions.

9. $\frac{3}{10}$%   10. $6\frac{3}{4}$%   11. $10\frac{1}{4}$%   12. $1\frac{4}{5}$%

Complete the following chart.

	Fraction	Decimal	Percent
13.	$\frac{7}{50}$	.14	
14.	$1\frac{2}{5}$		140%
15.	$\frac{1}{2000}$	.0005	
16.		.047	$4\frac{7}{10}$%

Study the following situations. Blacken the letter to the right that corresponds to the letter before the correct answer.

17. Two bakeries normally sell French bread for $1.00 a loaf. Store A is advertising the bread on sale at 20% off. Store B is offering $\frac{1}{4}$ off the usual price. At which store is the sale price lower?
    a. Store A   b. Store B   [a] [b]

18. Jones' Outlet is selling jeans for 30% off. Main Department Store is selling the same jeans for $\frac{1}{3}$ off. If both stores had the same original price, where could you get the better deal?
    a. Main Department Store   b. Jones' Outlet   [a] [b]

FRACTIONS AND PERCENTS

LESSON ONE: Changing Fractions to Percents

Instruction

Any fraction can be written as a percent. To write a fraction as a percent, follow these steps.

(1) Change the fraction to its decimal equivalent by dividing the numerator by the denominator.

(2) Move the decimal point two places to the right.

(3) Place the percent sign to the right of the number.

Example

$\frac{3}{5}$ = .6 $\longrightarrow$ .60. $\longrightarrow$ 60%

$\frac{1}{4}$ = .25 $\longrightarrow$ .25. $\longrightarrow$ 25%

$\frac{1}{20}$ = .05 $\longrightarrow$ .05. $\longrightarrow$ 5%

$\frac{1}{8}$ = .125 $\longrightarrow$ 12.5 $\longrightarrow$ 12.5%

$\frac{2}{3}$ = .66$\frac{2}{3}$ $\longrightarrow$ .66.$\frac{2}{3}$ $\longrightarrow$ 66$\frac{2}{3}$%

$3\frac{1}{3}$ = 3.33$\frac{1}{3}$ $\longrightarrow$ 3.33.$\frac{1}{3}$ $\longrightarrow$ 333$\frac{1}{3}$%

$8\frac{1}{4}$ = 8.25 $\longrightarrow$ 8.25. $\longrightarrow$ 825%

NOTE: The percents for $\frac{2}{3}$ and $3\frac{1}{3}$ have been written with fractions instead of decimals.

Directions

Write each fraction as its percent equivalent. If a fraction cannot be changed to a pure decimal (no fractional part), write the percent as a mixed fraction ($33\frac{1}{3}$%).

1. $\frac{4}{5}$ 2. $\frac{1}{5}$

3. $\frac{9}{10}$ 4. $\frac{7}{10}$

5. $\frac{5}{8}$ 6. $\frac{7}{8}$

7. $\frac{3}{20}$ 8. $\frac{17}{25}$

9. $7\frac{1}{3}$ 10. $\frac{7}{50}$

11. $\frac{1}{12}$ 12. $\frac{1}{9}$

13. $1\frac{1}{2}$ 14. $3\frac{3}{4}$

LESSON TWO: Changing Percents to Fractions

Instruction

Any percent can be written as a fraction. To write a percent as a fraction, follow these steps.
(1) Drop the percent sign.
(2) Place the number over 100.
(3) Reduce when possible.

Example

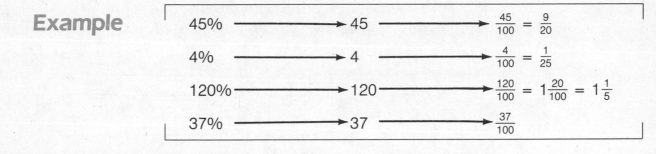

$45\% \longrightarrow 45 \longrightarrow \frac{45}{100} = \frac{9}{20}$

$4\% \longrightarrow 4 \longrightarrow \frac{4}{100} = \frac{1}{25}$

$120\% \longrightarrow 120 \longrightarrow \frac{120}{100} = 1\frac{20}{100} = 1\frac{1}{5}$

$37\% \longrightarrow 37 \longrightarrow \frac{37}{100}$

Exercise A

Write each percent as its fractional equivalent.

1. 50%
2. 28%
3. 6%
4. 3%
5. 250%
6. 10%
7. 1%
8. 175%
9. 25%
10. 13%
11. 325%
12. 9%

Exercise B

Write each percent as its fractional equivalent.

1. 60%
2. 72%
3. 2%
4. 7%
5. 210%
6. 18%
7. 4%
8. 125%
9. 75%
10. 19%

$\frac{1}{3}$ off

PRICE $12.00 $8.00

LESSON THREE: Changing Percents with Fractional Parts to Fractions

Instruction

When a percent has a fractional part, follow these steps to change it to a pure fraction.

(1) Change the fraction to a decimal by dividing the numerator by the denominator.

(2) Drop the percent sign.

(3) Move the decimal point two places to the left.

(4) Write the decimal as a fraction.

(5) Reduce if possible.

Example

$\frac{1}{4}\% \longrightarrow .25\% \longrightarrow .00.25 \longrightarrow \frac{25}{10000} \longrightarrow \frac{1}{400}$

$8\frac{1}{2}\% \longrightarrow 8.5\% \longrightarrow .08.5 \longrightarrow \frac{85}{1000} \longrightarrow \frac{17}{200}$

Exercise A

Write these percents as fractions.

1. $\frac{1}{2}\%$

2. $\frac{4}{5}\%$

3. $\frac{7}{10}\%$

4. $\frac{9}{10}\%$

5. $9\frac{1}{4}\%$

6. $8\frac{3}{4}\%$

Exercise B

Write these percents as fractions.

1. $1\frac{1}{2}\%$

2. $3\frac{1}{4}\%$

3. $\frac{2}{5}\%$

4. $\frac{3}{5}\%$

UNIT 9—REVIEW

Directions   Complete the following chart.

	Fraction	Decimal	Percent
1.	$\frac{9}{10}$		90%
2.	$\frac{3}{5}$	.6	
3.		.14	14%
4.	$\frac{1}{8}$	.125	
5.	$\frac{5}{80}$		$6\frac{1}{4}$%
6.	$\frac{1}{50}$	.02	
7.		.05	5%
8.	$\frac{7}{2000}$	.0035	
9.	$\frac{1}{200}$		$\frac{1}{2}$%
10.	$1\frac{1}{5}$	1.2	
11.	$\frac{1}{25}$		4%
12.		.02	2%
13.	$\frac{13}{50}$	.26	
14.	$\frac{1}{10}$		10%
15.		.47	47%
16.		.04	4%
17.		.004	$\frac{2}{5}$%
18.	$\frac{3}{50}$	.06	
19.	$\frac{27}{100}$		27%
20.	$\frac{1}{2000}$	.0005	
21.		.66	66%
22.	$\frac{3}{400}$		$\frac{3}{4}$%
23.		.72	72%
24.	$\frac{1}{400}$		$\frac{1}{4}$%

95

COMPARING DISCOUNTS

Stores often advertise sales in terms of a certain percent off regular prices. For example, a department store may advertise shirts at 20% off the regular price. Another store might advertise $\frac{1}{4}$ off its regular price. Discounts stated in terms of percents and fractions can be compared to determine which discount will save you the most money. See the example.

Example

You have a choice. Either you can change $\frac{1}{4}$ to a percent, or you can change 25% to a fraction.

$$\frac{1}{4} = .25 = 25\% \qquad\qquad 25\% = .25 = \frac{1}{4}$$

The discount here is equal. The original price of the merchandise would make the difference.

Directions   Study the following situations and answer the questions.

1. Store A and Store B are both going out of business. Store A offers all merchandise at 40% off. Store B offers all of its merchandise at $\frac{1}{3}$ off. Which store is offering the greater discount?

    ----

2. Ace Furniture Company is selling a dining room table and six chairs for 15% off. B & B Furniture Mart is selling the same set for $\frac{1}{5}$ off. If both stores had the same original price, where could you get the better deal?

    ----

3. Two tire stores are having sales on the same tire. Store One offers a 20% discount. Store Two's ad indicates when you buy one tire you get the second for half price. When buying a set of 4 tires, which store has the better price?

Courtesy The Goodyear Tire & Rubber Company

UNIT 10—PRETEST
PERCENT (PART = PERCENT × WHOLE)

Choose the answer which correctly completes each of the following statements. Circle the letter of your choice.

1. 4% of 400 is
   a. equal to 400.
   b. less than 400.
   c. greater than 400.

2. 101% of 36 is
   a. equal to 36.
   b. less than 36.
   c. greater than 36.

3. 87% of $\frac{1}{2}$ is
   a. equal to $\frac{1}{2}$.
   b. less than $\frac{1}{2}$.
   c. greater than $\frac{1}{2}$.

4. 100% of 110 is
   a. equal to 110.
   b. less than 110.
   c. greater than 110.

Solve the following problems.

5. What is $33\frac{1}{3}$% of 99? _____

6. 40% of 210 is _____.

7. 5% of 69 = _____.

8. $\frac{1}{2}$% of 28 is what? _____

Solve the following word problems. Blacken the letter to the right that corresponds to the letter before the correct answer.

9. There is a 5% sales tax on clothing. R.G. wants to buy four shirts costing $34 each. How much sales tax must he pay?
   a. $8.60   b. $6.85   c. $16.80   d. $6.80   [a] [b] [c] [d]

10. Marjorie's Boutique had a 20%-off sale. Yolanda found three dresses originally costing $56, $49, and $97. What did she pay for the three dresses?
    a. $161.60   b. $160.60   c. $165.60   d. $160   [a] [b] [c] [d]

11. Fred receives a commission of $\frac{1}{4}$% for each leasing agreement he arranges. He completes a transaction for $65,068. What will the amount of his commission be?
    a. $1,626.70   b. $162.67   c. $650   d. $16   [a] [b] [c] [d]

PERCENT (PART = PERCENT × WHOLE)

Understanding Percent

Instruction   When a coach tells an athlete, "Give me 100%," the coach wants the individual to give all the effort he or she can. When a student says, "I answered 100% of the problems on the test correctly," it means that he or she answered all of the problems correctly.

100% of a given number is that number.

Example

> 100% of 45 = 45   100% of 20 = 20   100% of 600 = 600

Daily situations usually involve some part which is less than 100% of the total amount.

Example

> 5% sales tax on a purchase
> $5\frac{1}{2}$% interest paid on a savings account
> 15% off the regular price

Percents less than 100% are less than the whole amount.

Example

> 80% of 45 = 36   5% of 20 = 1 $4\frac{1}{4}$% of 600 = 25.5

Exercise A   Choose the answer which correctly completes each of the following statements. Circle the letter of your choice.

1. 100% of 1 is
   a. equal to 1.
   b. less than 1.
   c. greater than 1.

2. 25% of 40 is
   a. equal to 40.
   b. less than 40.
   c. greater than 40.

3. 8% of 200 is
   a. equal to 200.
   b. less than 200.
   c. greater than 200.

4. 100% of $\frac{1}{2}$ is
   a. equal to $\frac{1}{2}$.
   b. less than $\frac{1}{2}$.
   c. greater than $\frac{1}{2}$.

5. 95% of 120 is
   a. equal to 120.
   b. less than 120.
   c. greater than 120.

6. 200% of 19 is
   a. equal to 19.
   b. less than 19.
   c. greater than 19.

7. 14% of 150 is
   a. equal to 150.
   b. less than 150.
   c. greater than 150.

8. 101% of 99 is
   a. equal to 99.
   b. less than 99.
   c. greater than 99.

9. 140% of 140 is
   a. equal to 140.
   b. less than 140.
   c. greater than 140.

10. $\frac{1}{2}$% of 50 is
    a. equal to 50.
    b. less than 50.
    c. greater than 50.

11. 2% of 67 is
    a. equal to 67.
    b. less than 67.
    c. greater than 67.

12. 205% of 4 is
    a. equal to 4.
    b. less than 4.
    c. greater than 4.

13. $8\frac{1}{2}$% of 18 is
    a. equal to 18.
    b. less than 18.
    c. greater than 18.

14. 300% of 100 is
    a. equal to 100.
    b. less than 100.
    c. greater than 100.

Exercise B   Choose the answer which correctly completes each of the following statements. Circle the letter of your choice.

1. 200% of 10 is
   a. equal to 10.
   b. less than 10.
   c. greater than 10.

2. 50% of $\frac{1}{4}$ is
   a. equal to $\frac{1}{4}$.
   b. less than $\frac{1}{4}$.
   c. greater than $\frac{1}{4}$.

3. 90% of 75 is
   a. equal to 75.
   b. less than 75.
   c. greater than 75.

4. $33\frac{1}{3}$% of $\frac{4}{5}$ is
   a. equal to $\frac{4}{5}$.
   b. less than $\frac{4}{5}$.
   c. greater than $\frac{4}{5}$.

5. 100% of 100 is
   a. equal to 100.
   b. less than 100.
   c. greater than 100.

6. 27% of 936 is
   a. equal to 936.
   b. less than 936.
   c. greater than 936.

LESSON TWO: Finding a Percent of a Number

Instruction

To find a percent of a number, multiply the number by the decimal equivalent of the percent.

part = percent × whole

Example

What is 25% of 40?

part = percent × whole

25% = .25   **First:** Change 25% to its decimal equivalent.

```
   .4 0
 × .2 5
   2 0 0
   8 0
 1 0.0 0
```

Second: Multiply 40 by .25

25% of 40 is 10.

The answer to this type of problem will not always be a whole number.

Example

What is 5% of 30?

part = percent × whole

5% = .05

```
    3 0
 × .0 5
  1.5 0
```

5% of 30 is 1.5.

Exercise A

Solve the following problems.

1. 40% of 440 is _____.

2. What is 20% of 85? _____

3. 3% of $72 = _____

4. 2% of $7,800 is what? _____

5. What is 25% of $64? _____

6. 50% of 60 is ....

7. $12\frac{1}{2}$% of 16 = ...

8. What is $33\frac{1}{3}$% of 45? ....

9. $\frac{1}{2}$% of 18 is what? ....

10. $8\frac{1}{2}$% of 70 is ....

11. What is 25% of $198? ....

12. $\frac{1}{4}$% of $1,600 = ...

13. $13\frac{1}{2}$% of 14 is ....

14. What is 70% of 120? ....

15. 80% of 82 = ...

16. 4.5% of 5 is ....

17. What is .5% of 250? ....

Exercise B   Solve the following problems.

1. 30% of 40 is ....

2. What is 72% of $425? ....

3. 25% of $165 = ...

4. 8.5% of $615 is ....

5. What is 12% of 84? ....

6. 75% of 120 = ...

7. $37\frac{1}{2}$% of 80 is what? ....

8. $66\frac{2}{3}$% of 15 is ....

9. $\frac{1}{4}$% of 20 = ...

10. $4\frac{1}{2}$% of 50 equals what? ....

11. What is 10% of $495? ....

12. $13\frac{1}{4}$% of 8 = ...

13. $12\frac{1}{2}$% of 6 is ....

14. 1.2% of 100 = ...

LESSON THREE: Solving Word Problems Involving Percents

Instruction

To solve word problems which require you to find a percent of a number, multiply the number by the decimal equivalent of the percent.

Example

There is a 5% sales tax on furniture. You buy $355 worth of furniture. How much sales tax must you pay?

THINK: What is 5% of $355?

SOLVE: 5% = .05

$$\begin{array}{r} \$3\,3\,5 \\ \times\ .0\,5 \\ \hline \$1\,6.7\,5 \end{array}$$

The sales tax will be $16.75.

Exercise A   Solve the following word problems.

1. A clerk sold $18,000 worth of clothes last year. He was paid a 15% commission. How much was his commission?   ....

2. To find a job for someone, an employment agency charges 7% of the client's pay the first month. If a medical technician takes a job paying $870 a month, what fee does she owe the employment agency?   ....

3. You are taking a test for your driver's license. The test has 25 questions, and the examiner tells you that in order to pass, you must answer 80% of the problems correctly. How many problems must you answer correctly to pass?   ....

4. A 3-pound jar of mixed nuts is labeled 60% peanuts. How many pounds of the mixed nuts are peanuts?   ....

SALE
EVERYTHING MUST GO
35% to 60% OFF!!

Exercise B   Solve these word problems.

1. Hezekiah made $20,750 last year and put 5% of that in a savings account. How much money did he save last year?

....

2. A utility company has a 6% charge for late payment. If your monthly bill is $75.50, what is the charge for late payment?

....

3. When you hire a lawyer to settle a $15,000 estate, she says that her fee will be 12% of the amount to be settled. What is the lawyer's fee?

....

4. Mrs. Hodge is paid $575 a month at her part-time job. Her boss says she is going to get an 8% raise. How much will the raise be?

....

5. A newsstand grossed $17,000 last year. Sales are expected to drop 9%. How much less should sales be this year than last year?

....

6. The number of people in the city covered by a TV station is 350,000. If channel 7 reaches 125,000 of those people, what is its share (percentage) of the population?

....

7. Capital City's population is 1980 was 475,000. If its population in 1990 is 39% larger, what will the population be?

....

103

Solving Problems Involving Discounts

Instruction

A **discount** is the amount of reduction made from the whole amount of something. Sometimes a problem asks for the amount of the discount. Sometimes a problem asks for the amount left after the discount.

Example

An appliance store advertises 15% off all refrigerators. The price of the refrigerator you have been looking at was $875. How much will you save by buying it on sale? What is the sale price?

Step 1

THINK: 15% of $875 is ....?
15% of $875 is $131.25.
This is the amount of discount.

$$\begin{array}{r} \$875 \\ \times\ .15 \\ \hline 4375 \\ 875 \\ \hline \$131.25 \end{array}$$

Step 2

$$\begin{array}{r} \$875.00 \quad \text{old price} \\ -\ 131.25 \quad \text{discount} \\ \hline \$743.75 \quad \text{sale price} \end{array}$$

The sale price is $743.75.

Exercise A Solve the following word problems.

1. You go to a store that is having a 25%-off sale and find a dress which regularly sells for $72. What will you have to pay for the dress on sale? ....

2. Mario bought a $56 pair of shoes and a $28 pair of slacks at a 15%-off sale. How much was his total bill? ....

3. Marjorie's car needs new tires. The tires regularly sell for $75.95 each. If she buys 4 tires, she will get a 10% discount. What will her total bill be for four tires at the discount price? ....

Exercise B   Solve the following word problems.

1. Jim's father said that he would help Jim buy a motorcycle. He told Jim that he would pay 25% of the cost of the cycle. Jim found a used motorcycle he wanted for $325. How much must Jim pay?   ....

2. Terry is buying a used car for $1,200. He pays 40% down and will pay the remainder in installments. What amount does he owe in installments?   ....

3. Brenda takes home $620 each month. She decides to save 5% of each check. How much money will she have to spend each month?   ....

4. An insurance company offers a $1\frac{1}{2}$% discount to policyholders who pay their premium in one payment at the first of the year. Bill Taylor's premium is $175. How much would he pay in one lump sum at the beginning of the year?   ....

5. Chuck Balding is a salesperson. He is paid a commission of 2% of all sales he makes. His company deducts 6.7% for Social Security and 20% for Federal Income Tax. If Chuck sells $23,568, how much will his check be?   ....

105

UNIT 10—REVIEW

Directions   Solve the following problems.

1. 25% of 800 is _____.

2. What is 40% of 120? _____

3. 18% of $35 = _____

4. $1\frac{1}{2}$% of $16 is what? _____

5. 5% of $1.60 is _____.

Solve the following problems. Blacken the letter to the right that corresponds to the letter before each correct answer.

6. A dishwasher listed at $480 is marked 20% off at a sale. What is the sale price?
   a. $384   b. $434   c. $386   d. $385   [a] [b] [c] [d]

7. Doyle's gross salary is $820 a month. The following deductions are made from his check: 11% federal tax, 5% state tax, and $7.50 for union dues. What is his net (take-home) salary for a month?
   a. $630.80   b. $683.80   c. $681.30   d. $681.80   [a] [b] [c] [d]

8. At a 10%-off sale you purchase a television set originally marked $595. If a 6% sales tax is charged, what do you pay for the television?
   a. $666.73   b. $567.63   c. $586.60   d. $565   [a] [b] [c] [d]

9. A real estate agent sells your home for $78,500. How much was her commission if she received 6% of the selling price?
   a. $4,710   b. $3,750   c. $4,770   d. $4,570   [a] [b] [c] [d]

10. Felton's Department Store mails a monthly statement to its customers. One bill showed a $17.57 purchase and a $1\frac{1}{2}$% finance charge. What is the total bill?
    a. $24.67   b. $17.87   c. $16.83   d. $17.83   [a] [b] [c] [d]

Directions

Solve the following problems.

1. 40% of 50 is ....

2. What is 48% of 700? ....

3. 16% of $530 = ...

4. $62\frac{1}{2}$% of $320 is ....

5. A local theater offers a 10% discount to students. What do students pay for tickets which normally sell for $2.50? ....

6. In June, city workers received a 5% salary increase. A police officer whose annual salary was $18,750 will now receive what amount per month? ....

7. Morgan City has a 4% sales tax on purchases made in the city. Mrs. Clowers pays $86.45 for clothes for her son and $24.80 for cosmetics for herself. What is the total sales tax on her purchases? ....

8. Ted Simpson filed a tax return on a taxable income of $23,800. The tax was $4,380 plus 32% of the amount over $20,000. What total income tax did he pay? ....

9. The local hardware store is having a sale on paint. Gallons with a regular price of $9.99 are 10% off. If Jane bought 6 gallons and had to pay sales tax of 5%, what would be the total amount she would pay? ....

WARRANTIES

Any **warranty** on purchased merchandise should be understood by the buyer at the time the purchase is made. The terms of warranties differ. Basically, a warranty is a written guarantee on a product. It specifies the seller's responsibility for the repair or replacement of defective items. There are both **full** warranties and **limited** warranties.

Under a full warranty, you are not charged for repair or replacement costs during the length of the warranty. With a limited warranty, you must pay part of the repair or replacement costs.

Look at the tire warranty below. Read it carefully. Then answer the questions. Use the price chart when figuring costs of replacement tires.

GALAXY TIRE WARRANTY
Full 24-month Warranty Plus Limited 24-month Warranty

Length of Warranty	Free Replacement	Replacement Charge	
		50%	75%
48 Months	1–24 Months	25–36 Months	37–48 Months

Galaxy tires are warranted for 48 months from the date of original purchase against defects in materials and workmanship, road hazards (except repairable punctures), and tread wear. **What We Will Do:** Galaxy will replace the tire with a new, identical tire, if available, or with a new tire of equal or better quality without charge to the customer during the first 24 months of the warranty period. During the next 12 months of the warranty period, the customer will be charged 50% of the current price of the replacement tire. During the last 12 months of the warranty period, the customer will be charged 75% of the current price of the replacement tire. This warranty does not cover tires used for commercial service, simple repairable punctures, or failure due to fire, theft, accident, and chain or mounting damage.

Steel-Belted Radial Tires

Size	Blackwalls	Whitewalls
155/80R13	$52	$55
185/80R13	$58.50	$61.50
195/75R14	$67	$70
205/75R14	$73.50	$76.50
215/75R14	$80.50	$83.50
215/75R15	$86.50	$89.50
225/75R15	$84.50	$87.50
235/75R15	$88.50	$90.50
245/75R15	$90.50	$93.50
255/75R15	$99.50	$102.50

1. Full warranty on Galaxy tires lasts how many months? ....

2. Defective tires returned in the third year of ownership may be replaced at what percent of the current selling price? ....

3. You return one defective 205/75R14 blackwall tire to the company 27 months after the date of purchase. The company will sell you a replacement at what cost? ....

4. If you return two defective 155/80R13 whitewall tires within 42 months from the date of purchase, what will the total replacement charge be for two tires? ....

5. Linda Johnson bought two 215/75R14 blackwall tires 50 months ago. One of them is defective. How much will the charge be for the replacement tire? ....

6. Tom Hunt has a set of 195/75R14 Galaxy blackwall tires on his car. He purchased them 20 months ago. This morning he found a small nail in one of the tires. How much will a replacement cost him? (The nail hole can be repaired.) ....

UNIT 11—PRETEST
PERCENT (WHOLE = PART ÷ PERCENT)

Solve the following problems.

1. 12 is 20% of _____.

2. 65% of _____ is 65.

3. 27 is 25% of what number? _____

4. $62\frac{1}{2}$% of _____ is 30.

5. 6 is 15% of _____.

6. 200 is 80% of what number? _____

7. 30% of _____ is 9.6.

8. 2.8 is 10% of _____.

Solve the following word problems. Blacken the letter to the right that corresponds to the letter before each correct answer.

9. $16 is deducted from each of Marilyn's paychecks for uniforms. This is 2% of her salary. What is her salary?
   a. $800   b. $1600   c. $875   d. $470   ⓐ ⓑ ⓒ ⓓ

10. Joan's 2% commission on sales amounted to $37 this week. What were her total sales?
    a. $1,250   b. $1,500   c. $1,850   d. $1,600   ⓐ ⓑ ⓒ ⓓ

11. 2% of the employees in one plant contributed 1% of their earnings to the United Way. These 14 employees had the deductions made from their monthly checks. How many people work in this particular plant?
    a. 650   b. 200   c. 600   d. 700   ⓐ ⓑ ⓒ ⓓ

12. Residents in one state pay a $12\frac{1}{2}$% property tax on the appraised value of their property. What would the value of ten acres be if the owner paid $870 in property taxes?
    a. $6,860   b. $6,960   c. $6,690   d. $6,100   ⓐ ⓑ ⓒ ⓓ

13. 5% of Oscar's salary is deducted for his retirement plan. What is his annual salary if his retirement deduction is $115 a month?
    a. $27,600   b. $26,700   c. $25,450   d. $20,600   ⓐ ⓑ ⓒ ⓓ

PERCENT (WHOLE = PART ÷ PERCENT)

LESSON ONE:
Finding a Number from Its Part

Instruction

To find a number when a part of it is known, divide the part by the decimal equivalent of the percent.

whole = part ÷ percent

Example

10 is 40% of what number?

40% = .40   **First:** Change 40% to its decimal equivalent.

$$.40\overline{)10.00}$$
$$\begin{array}{r} 25. \\ \underline{-80} \\ 200 \\ \underline{-200} \end{array}$$

Second: Divide 10 by .40.

10 is 40% of 25.

The numbers in this type of problem will not always be whole numbers.

Example

6.348 is 34.5% of what number?

34.5% = .345

$$.345\overline{)6.3480}$$
$$\begin{array}{r} 18.4 \\ \underline{-345} \\ 2898 \\ \underline{-2760} \\ 1380 \\ \underline{-1380} \end{array}$$

6.348 is 34.5% of 18.4.

Exercise A

Solve the following problems.

1. 40 is 25% of what number? _____

2. 10 is 25% of _____.

3. 25% of _____ is 18.

4. 25% of _____ is 1,620.

5. $62\frac{1}{2}$% of _____ is 15.

6. $3.75 is 25% of _____.

7. $9.45 is 45% of _____.

8. 64% of _____ is $2.88.

9. $6.75 is $33\frac{1}{3}$% of _____.

10. 6% of _____ is 54.

11. $12.60 is 90% of _____.

12. 30% of _____ is $603.

13. 25% of _____ is 49.5.

14. 840 is 60% of _____.

Exercise B   Solve the following problems.

1. 16 is 20% of what number? _____

2. 6 is 5% of _____.

3. 70% of _____ is 14.

4. 75% of _____ is 264.

5. 1.7 is 10% of _____.

6. 100 is 40% of _____.

7. $4.50 is 75% of _____.

8. 80% of _____ is $36.

9. 9 is $33\frac{1}{3}$% of _____.

10. 8% of _____ is 64.

11. 40% of _____ is 160.

12. $24 is 20% of _____.

13. 10% of _____ is 10.

14. 15% of _____ is 4.8.

Solving Word Problems Involving Percent

Instruction

To solve word problems which require you to find a number when a part and a percent of that number are known, divide the part by the decimal equivalent of the percent.

Example

> Yolanda saves $25 each week from her earnings. This amount represents 5% of her earnings. How much money does she earn each week?
>
> THINK: $25 is only part of a larger amount. $25 is 5% of
>
> ----------------------.
>
> SOLVE: 5% = .05
>
> $$.05 \overline{)\,\$25.00} \quad \begin{array}{r} \$500. \\ -25 \\ \hline 0 \\ -0 \\ \hline 0 \\ -0 \end{array}$$
>
> She earns $500 each week.

Exercise A   Solve the following word problems.

1. Paul and Rhonda bought a new house. They made a down payment of $8,800. This represented a 10% down payment. What was the cost of the house?   ----------------------

2. 5% of this month's utility bill is for a $3 fuel adjustment charge. How much was the utility bill?   ----------------------

Exercise B   Solve the following word problems.

1. Al works 40 hours a week. Each day he makes $56. This is 20% of a week's pay. How much does Al make each week?   ----------------------

2. Discount Drug Store sells cosmetics at a 10% discount. Lipstick there costs $2.07. What is the regular price of the lipstick?   ----------------------

UNIT 11—REVIEW

Directions   Solve the following problems.

1. 60% of _____ is 3.

2. $.08 is 5% of _____ .

3. 80% of _____ is $12.80.

4. 22 is 16% of _____ .

5. 10 is $12\frac{1}{2}$% of _____ .

Solve the following problems. Blacken the letter to the right that corresponds to the letter before each correct answer.

6. In a recent election, the winner received 2,457 votes, or 54% of all votes cast. How many voters participated in the election?
   a. 4,450   b. 4,550   c. 4,250   d. 4,650   ⓐ ⓑ ⓒ ⓓ

7. You buy a coffee maker for $34. It was on sale for 15% off. What was the original price of the appliance?
   a. $45   b. $60   c. $40   d. $55   ⓐ ⓑ ⓒ ⓓ

8. Walter stopped by the convenience store for a can of cleanser. The 5% sales tax on his purchase was $.04. What was the amount of his purchase?
   a. $.80   b. $8.00   c. $1.80   d. $.08   ⓐ ⓑ ⓒ ⓓ

9. Lloyd sold his bicycle for $48. That was 25% less than what he had paid for it. How much had the bicycle cost Lloyd?
   a. $64   b. $46   c. $67   d. $47   ⓐ ⓑ ⓒ ⓓ

10. Wanda bought a dress at 25% off and paid $24 for it. What was the original price of the dress?
    a. $18   b. $30   c. $27   d. $32   ⓐ ⓑ ⓒ ⓓ

11. 15% of Joe Walker's annual salary goes for income tax. What was his annual salary last year if his income tax was $4,575?
    a. $30,500   b. $32,000   c. $28,500   d. $30,000   ⓐ ⓑ ⓒ ⓓ

12. At a 12%-off sale, you find a lamp marked $22.88. What was the original price of the lamp?
    a. $28   b. $27   c. $29   d. $26   ⓐ ⓑ ⓒ ⓓ

EMPLOYMENT AGENCIES

State and private employment agencies are in business to help people find jobs. When a state employment agency helps secure jobs for people, there is no charge. However, private employment agencies charge for their services. The fees vary from agency to agency, although their rates are regulated by state and federal laws. The Better Business Bureau offers the following suggestions about using private employment agencies.

- Use an employment agency only if you are fully aware of and able to meet the terms of the contract governing fees.

- Be certain that the employment agency which you select is reputable. The Better Business Bureau maintains files on most employment agencies. Call for a responsibility report.

- If an agent gives oral promises of conditions contrary to those in the contract, be sure that these are in writing and signed by the agent.

- Read and fully understand the contract before signing it.

- Be honest with the agency. If you withhold adverse information about your background, the truth will come out eventually and may be more damaging than if you had been honest in the beginning.

- If you discover that an agency has sent you to a company which has not requested applicants or listed an opening with the agency, have no further dealings with the agency. The same is true if the agency sends you to jobs for which you are not qualified.

- Cooperate with the agency. It is to your advantage. When you are sent on an interview, let the employment agency know the results immediately.

- Take the time to properly prepare for your interview. An employment agency's efforts are useless if you fail to convince an employer that you are the best person for the job.

- Give careful consideration to the acceptance of a position. An agency's responsibility is to get you interviews. It is your responsibility to accept the job which is best for you.

1. Suppose you find and accept a job through a private employment agency. Your contract with the agency stipulates that 9 percent of your first year's annual gross pay will be payable to the agency for their services. The agency bills you for $1,593. What is your annual salary?

   ----

2. Sue Cyler accepts a position with Allied Industries. Summit Employment Agency, which secured the job for her, has a service charge of 5% of her gross yearly earnings. If Sue pays the agency $900 for services, what is her annual salary?

   ----

3. George Daughty signed a contract with an employment agency which found him a job. The contract stated that George would pay the agency 12% of his gross yearly earnings. The fee is $1,504.32. What is George's annual salary?

   ----

4. Don Forbes accepted a position after an employment agency had gotten him the interview. The contract stipulated that he pay 20% of the first two-months' salary. The fee was $500. What was Don's annual salary?

   ----

5. Tony took an employment test and missed 17% of the problems. He worked 166 problems correctly. How many problems were on the test?

UNIT 12—PRETEST
PERCENT (PERCENT = PART ÷ WHOLE)

Solve the following problems.

1. 4 is ....% of 5.

2. 56.25 is ....% of 75.

3. What percent of 176 is 88? ....

4. What percent of 252 is .63? ....

Solve the following word problems. Blacken the letter to the right that corresponds to the letter of each correct answer.

5. One car dealer will sell you the car you want for $6,400. Another dealer will sell the car for $5,600. The lower price is what percent less than $6,400?
   a. 11%   b. 11.5%   c. 12%   d. 12.5%   [a] [b] [c] [d]

6. The total deductions made from Fred Lewis' monthly check of $1,066 amount to $160. $11.20 of the deductions is for union dues. What percent of the total deductions goes for union dues?
   a. 6%   b. 7%   c. 7½%   d. 5%   [a] [b] [c] [d]

Read and solve the following. Blacken the letter to the right that corresponds to the letter of each correct answer.

7. A pair of $64 shoes are on sale for $51.20. What is the percent of savings?
   a. 20%   b. 25%   c. 19%   d. 12%   [a] [b] [c] [d]

8. Maribel bought a car for $15,000 and sold it for $9,000. What percent did the value of the car depreciate?
   a. 35%   b. 38%   c. 42%   d. 40%   [a] [b] [c] [d]

9. Cora's salary increased from $28,000 to $30,520 a year. What is the percent of increase?
   a. 7%   b. 9%   c. 8%   d. 6%   [a] [b] [c] [d]

10. City bus fare has increased from $.50 to $.60. What is the percent of increase?
    a. 16%   b. 25%   c. 20%   d. 19%   [a] [b] [c] [d]

PERCENT (PERCENT = PART ÷ WHOLE)

LESSON ONE: **Finding the Percent One Number Is of Another Number**

Instruction
To find the percent one number is of another number, divide the part by the whole amount.

$$percent = part \div whole$$

Example

4 is what percent of 5?

$$\begin{array}{r} .8 = 80\% \\ 5\overline{)4.0} \\ -40 \\ \hline \end{array}$$

Divide 4 by 5. Move the decimal point two places to the right. Write the percent sign.

4 is 80% of 5.

Sometimes the percent will be greater than 100% or less than 1%.

Example

8 is what percent of 4?

$$\begin{array}{r} 2 = 200\% \\ 4\overline{)8} \\ -8 \\ \hline \end{array}$$

8 is 200% of 4.

.08 is what percent of 16?

$$\begin{array}{r} .005 = .5\% \text{ or } \frac{1}{2}\% \\ 16\overline{).080} \\ -80 \\ \hline \end{array}$$

.08 is $\frac{1}{2}$% of 16.

Exercise A Solve the following problems.

1. 15 is _____% of 20.

2. 16 is _____% of 64.

3. 3 is what percent of 60? _____

4. 1 is what percent of 3? _____

5. What % of 50 is 35? _____

6. What percent of 350 is 42? ....

7. 5 is ....% of 12.

8. 33 is ....% of 200.

9. What percent of 96 is 12? ....

10. 2 is ....% of 125.

Exercise B   Solve the following problems.

1. 6 is ....% of 40.

2. 17 is ....% of 85.

3. 3 is what percent of 4? ....

4. 7 is what percent of 8? ....

5. What % of 52 is 13? ....

6. What percent of 65 is 39? ....

7. 9 is ....% of 54.

8. 14 is ....% of 700.

9. What percent of 125 is 75? ....

10. 1 is ....% of 200.

11. 16 is ....% of 24.

12. What percent of 800 is 8? ....

13. 2 is what percent of 3? ....

14. 62 is ....% of 124.

15. 99 is ....% of 1,980.

16. What percent of 66 is 20?

....

17. 7 is what percent of 10?

....

18. What percent of 250 is 75?

....

```
22/05/84  39/3        0640    500402

MDSE             40430113      16.00
SUBTOTAL                       16.00
5% SALES TAX                     .80
CASH TENDERED                  20.00
CHANGE DUE                      3.20

      1 CASH SALE              16.80


*  PLEASE PRESENT THIS RECEIPT  * *
*        IN CASE OF INQUIRY       * *
*                                  *
```

LESSON TWO: Solving Word Problems Involving Percent

Instruction

To solve word problems which require you to find what percent one number is of another number, divide the part by the whole amount.

Example

> Madge wanted to buy a food processor which cost $280 regularly. By paying cash, she could buy it for $266. What percent would she save by paying cash?
>
> THINK: savings = regular price − cash price
>
> THINK: percent = part ÷ whole
>   percent = savings ÷ regular price
>
> SOLVE: savings = $\begin{array}{r} \$280 \\ -266 \\ \hline \$14 \end{array}$ percent = $\begin{array}{r} .05 = 5\% \\ 280\overline{)14.00} \\ -1400 \\ \hline \end{array}$
>
> Madge saved 5% by paying cash.

Exercise A   Solve the following word problems.

1. One store advertised its $150 vacuum cleaner on sale for $127.50. What percent could be saved by purchasing the vacuum cleaner on sale?   ------------------

2. 275 students attend Lincoln High School. Of these students, 110 are girls and 165 are boys. What percent of the enrollment is girls?   ------------------

Exercise B   Solve the following word problems.

1. Jane Rowan spent $50 on groceries this week. $14 of this amount was for meat. What percent of the grocery bill was for items other than meat?   ------------------

2. During a 40-hour work week, Elise spends 13 hours traveling, 20 hours selling, and 7 hours writing reports. Traveling consumes what percent of her work week?   ------------------

LESSON THREE: Finding Percent of Increases or Decreases

Instruction

You can find the **percent of increase** by finding the amount of increase (gain) and dividing it by the original amount.

percent increase = gain ÷ original amount

You can find the **percent of decrease** by finding the amount of decrease (loss) and dividing it by the original amount.

percent decrease = loss ÷ original amount

Example

The price of a dozen eggs increased from $1.20 to $1.50 per dozen. What is the percent of increase?

THINK: percent increase = gain ÷ original amount

SOLVE:
```
gain =    7 5 ¢          percent = 60)15.00
        - 6 0 ¢                    - 1 2 0
          1 5 ¢                      3 0 0
                                   - 3 0 0
```
$$\frac{.25}{60)15.00} = 25\%$$

The percent of increase is 25%.

Exercise A Solve the following word problems.

1. Jeff bought a bicycle for $120. He used the bicycle for 6 months and sold it for $90. What percent did the value of the bike depreciate (decrease in value)? ....

2. The Andersons' rent increased from $300 a month to $400 a month. What was the percent of increase? ....

3. Rita bought a $50 raincoat for $40. What percent did she save? ....

Exercise B Solve these word problems.

1. Madeline's car insurance rose from $320 annuallly to $336 annually. What percent did her insurance increase? ....

2. What percent of increase did workers receive if their hourly pay rate changed from $6 to $7.20 an hour? ....

121

UNIT 12 — REVIEW

Directions   Solve the following problems.

1. _____% of $6 is $1.50.

2. 16 is what percent of 64? _____

3. 14 is _____% of 700?

4. What percent of 5 is 3? _____

5. 1 is _____% of 8.

Solve the following problems. Blacken the letter to the right that corresponds to the letter of each correct answer.

6. The Thompsons are buying a $60,000 home. Their down payment is $12,000. What percent did they pay down?
   a. 18%   b. 20%   c. 16%   d. 25%

   a b c d

7. Ralph bought a tie on sale for $5.40. It sold originally for $6. What percent did he save?
   a. 8%   b. 9%   c. 10%   d. 11%

   a b c d

8. Mrs. Johnson received an annual salary increase of $1,400. If her previous salary was $14,000, what percent did her salary increase?
   a. $9\frac{1}{2}$%   b. 11%   c. 15%   d. 10%

   a b c d

9. There were 50 questions on Billy's science test. He missed 17 questions. What percent did he get correct?
   a. 65%   b. 60%   c. 66%   d. 70%

   a b c d

10. Ms. Ling's Social Security check increased from $200 to $233 a month. What percent of increase does this reflect?
    a. 16.5%   b. 17%   c. 16%   d. 15%

    a b c d

11. We are driving to a football game 240 miles away. When we have traveled 180 miles, what percent of the distance remains?
    a. 25%   b. 20%   c. 80%   d. 75%

    a b c d

12. Paul works on commission. His commission for the month of September was $720. His commission in October was $576. This was what percent less than his September commission?
    a. 16%   b. 20%   c. 25%   d. 18%

    a b c d

HEALTH INSURANCE

Almost 90% of all Americans have some form of health insurance. Health, or medical, insurance may be purchased individually or as part of a group. Many employers offer group medical insurance as one of their employee benefits. The employer sometimes pays the full amount of the **premium**. Often, the employer and the employee share the cost of the insurance. Usually, group insurance costs you less than the same coverage would cost if you bought it individually.

The main types of medical insurance plans provide the following coverages.

Hospital Expense—Hospitalization insurance provides for a certain amount of hospital charges such as room, board, and other services to be paid for. The policy usually will pay for the use of operating rooms, anesthesia, laboratory and X-ray examinations, and drugs and other services provided by the hospital. The policy will limit coverage to a certain number of days per hospital visit or per year.

Surgical Expenses—If the services of a surgeon are needed, the insurance will usually pay his or her fee according to a set schedule of fees which covers all or part of the cost of each type of operation.

Some policies provide for full payment of **usual, customary, and reasonable** surgical charges. There is no fixed schedule of fees. Instead, the insurance company will pay the full amount as long as it is the usual fee the doctor charges the majority of patients for the same services, it is no more than the fee customarily charged by most doctors in the same locality, and it is not unreasonable for the services performed.

Medical expenses—Coverage for medical expenses provides payments for all or part of the charges made by doctors other than surgeons for hospital calls. Some policies also provide coverage for medical treatment at home and/or in the doctor's office.

Major Medical Expenses—A major medical policy provides for payment of a set percentage of your medical expenses above a certain amount, the **deductible**. Usually, the policy will pay for 80% of the covered expenses above the deductible amount. The amount of the deductible varies. There is a maximum amount which the insurance will pay under a major medical policy.

You should be familiar with the terms of your health insurance plan, if you have one. Insurance companies will not pay for medical expenses unless you, your doctor, or your hospital file a claim. You should be familiar with the following items in your health insurance policy.

- What medical expenses are covered and what medical expenses are not covered?

- Does the policy pay for surgical procedures according to a schedule, or does it pay usual, customary, and reasonable charges?

- Do you have major medical coverage in addition to hospitalization and medical expense coverage?

- How much is your deductible on your major medical insurance?

- Does your medical insurance cover all members of your family or only yourself?

If you do not understand your policy or have any questions concerning how to file claims, see the insurance company's agent or the insurance or personnel manager where you work.

Directions Solve the following problems involving insurance.

1. Mary Lou's medical insurance pays 80% of covered illnesses. It includes a $100 deductible as well. Her first claim is a doctor bill for $280. How much will she be paid by the insurance company for her claim?

----

2. Jack Fritch has medical insurance where he works. His company pays for his coverage. He must pay for his wife's insurance through payroll deduction. The monthly premium for her coverage is $168. How much will be deducted from Jack's semi-monthly paycheck?

----

3. Juan Gomez has dental insurance that pays 80% of routine care and 60% of major dental work. A $50 deductible is a part of the policy. Juan's dentist charged him $40 for an examination in January and $100 for a new crown on one of his teeth in May. How much will he be paid when he files his claim?

FINAL REVIEW

Blacken the letter to the right that corresponds to the letter before each correct answer.

1. .7 =   a. $\frac{7}{100}$ b. $\frac{7}{10}$ c. $\frac{7}{1}$ d. $\frac{7}{1000}$ [a] [b] [c] [d]

2. .4204 =   a. $\frac{4204}{1000}$ b. $\frac{4204}{100}$ c. $\frac{4204}{10000}$ d. 4,204   [a] [b] [c] [d]

3. .14 =   a. $\frac{14}{10}$ b. $\frac{14}{1000}$ c. $\frac{14}{100}$ d. $\frac{14}{10000}$ [a] [b] [c] [d]

4. .201 =   a. $\frac{201}{100}$ b. $\frac{201}{10}$ c. $\frac{201}{10000}$ d. $\frac{201}{1000}$ [a] [b] [c] [d]

5. 7 tenths =   a. .07   b. $\frac{7}{10}$ c. .007   d. $\frac{7}{100}$ [a] [b] [c] [d]

6. 4 thousandths =   [a] [b] [c] [d]
   a. .004   b. .0004   c. $\frac{4}{100}$ d. $\frac{4}{10}$

7. seven hundred forty-six and two ten-thousandths =   [a] [b] [c] [d]
   a. 746.2   b. 746.0002   c. $7426\frac{2}{000}$
   d. $746\frac{2}{100}$

8. two thousand four and seven hundredths =   [a] [b] [c] [d]
   a. $204\frac{7}{100}$ b. 2,004.7   c. $2,004\frac{7}{10}$
   d. 2,004.07

9. five dollars and eighteen cents =   [a] [b] [c] [d]
   a. $5.18   b. 5.18   c. $5\frac{18}{10}$ d. $51.8

10. twenty-two dollars and twenty-five cents =   [a] [b] [c] [d]
    a. 2225   b. $22.0   c. $22.25
    d. $22\frac{25}{100}$¢

11. sixty-eight dollars and seventy cents =   [a] [b] [c] [d]
    a. $68.70   b. $68\frac{700}{100}$ c. 6870
    d. 68.75

12. eight dollars and twelve cents =   [a] [b] [c] [d]
    a. $812   b. $8\frac{12}{100}$¢   c. $8.12   d. $\frac{812}{1000}$

13. .067 =   a. .0067   b. .0670   c. 6.7   d. 60.7   [a] [b] [c] [d]

14. .662 =   a. 6.62   b. 66.2   c. .66200   [a] [b] [c] [d]
    d. 0662

15. 6. =   a. 6.00   b. .6   c. 60.0   d. 60.00   [a] [b] [c] [d]

16. 5.23 =   a. .523   b. 52.3   c. .0523   d. 5.230   [a] [b] [c] [d]

17. 5,876.06 + 324.61 =   [a] [b] [c] [d]
    a. 6,201.21   b. 6,200.67   c. 9,122.16
    d. 9,022.16

18. $16 + $255.11 + $9 =   [a] [b] [c] [d]
        a. $2,801.10   b. $1,315.11
        c. $280.11   d. $131,511.00

19. 38.001 + 654 + 4.02 =   [a] [b] [c] [d]
        a. 1436.01   b. .696021   c. 696.021
        d. 14.3601

20. $87.50 + $6.93 + $3 =   [a] [b] [c] [d]
        a. $186.80   b. $97.43   c. $94.73
        d. $94.37

21. 5.0007 + .013 + 8 =   [a] [b] [c] [d]
        a. 13.0137   b. 13.1307   c. 130.137
        d. 131.307

22. 4.6063 − 2.5192 =   [a] [b] [c] [d]
        a. 20,871   b. 7.1255   c. 712.55
        d. 2.0871

23. $2.98 − $1.99 =   [a] [b] [c] [d]
        a. $1.00   b. $.99   c. $1.99
        d. $1.98

24. $5.47 − $3.85 =   [a] [b] [c] [d]
        a. $1.62   b. $10.32   c. $16.62
        d. $103.20

25. .9426 − .0294 =   [a] [b] [c] [d]
        a. .972   b. 9.132   c. .9132
        d. .6486

26. $80 − $49.95 =   [a] [b] [c] [d]
        a. $30.50   b. $28.05   c. $29.05
        d. $30.05

27. 836.6 − .549 =   [a] [b] [c] [d]
        a. 837.149   b. 836.051   c. 1,385.6
        d. 837,149

28. 24.031 − 19.2 =   [a] [b] [c] [d]
        a. 4,831   b. 4.831   c. 42.231
        d. 42,831

29. $101.09 − $25 =   [a] [b] [c] [d]
        a. $126.09   b. $127.09   c. $76.09
        d. $77.09

30. .217   a. .651   b. .0651   c. 6.51   d. 65.1   [a] [b] [c] [d]
   × .3

    .744   a. .13392   b. .013392   c. 13.392   [a] [b] [c] [d]
31. × .18   d. 1.3392

32.   .3   a. .99   b. .09   c. 9   d. .9   [a] [b] [c] [d]
    $\times$ .3

33. $2\overline{)3.108}$ a. 1.554   b. .1554   c. .01554   [a] [b] [c] [d]
    d. 15.554

34. $12\overline{)4.812}$ a. .0401   b. .401   c. 4.01   d. 401   [a] [b] [c] [d]

35. $21\overline{)1.260}$ a. .6   b. .006   c. 6   d. .06   [a] [b] [c] [d]

36. $33\overline{)693.33}$ a. .2101   b. 2.101   c. 21.01   [a] [b] [c] [d]
    d. 210.1

37. $4\overline{)2}$ a. .05   b. 5   c. .5   d. .55   [a] [b] [c] [d]

38. $5\overline{)4.15}$ a. .083   b. .83   c. .803   d. 8.3   [a] [b] [c] [d]

39. $18\overline{)1.08}$ a. .6   b. 6   c. .006   d. .06   [a] [b] [c] [d]

40. $4.5\overline{)4.6215}$ a. .1027   b. 10.27   c. 1.027   [a] [b] [c] [d]
    d. 102.7

41. $1.3\overline{)11.7}$ a. 9.1   b. .009   c. .09   d. 9   [a] [b] [c] [d]

42. $21.1\overline{)10.55}$ a. .05   b. .5   c. 5   d. 50   [a] [b] [c] [d]

43. $.27\overline{).0162}$ a. .06   b. 6   c. .6   d. .0006   [a] [b] [c] [d]

44. $3.4\overline{)1.156}$ a. 3.4   b. .34   c. .034   d. 34   [a] [b] [c] [d]

45. $.6\overline{)606}$ a. 1.010   b. 1,010   c. 101.0   [a] [b] [c] [d]
    d. 10,100

46. $.001\overline{)613}$ a. 613,000   b. 613.000   c. 613.00   [a] [b] [c] [d]
    d. 61,300

47. $4.2\overline{)4.70}$ a. 112.0   b. 1.120   c. 11,120   [a] [b] [c] [d]
    d. 11,120.0

48. $7.9\overline{)17.222}$ a. .218   b. .0218   c. .00218   [a] [b] [c] [d]
    d. 2.18

49. $110\overline{)21.9}$ a. .2   b. 20   c. 2   d. 22   [a] [b] [c] [d]

50. $9\overline{)70}$ a. 778   b. 77.8   c. 7.78   d. .778   [a] [b] [c] [d]

51. $.71\overline{)14.526}$ a. 20.46   b. 2.046   c. .2046   [a] [b] [c] [d]
    d. .02046

52. $306\overline{)43.67}$ a. 14   b. 1.4   c. .014   d. .14   [a] [b] [c] [d]

53. $\frac{1}{16} =$ a. $.06\frac{1}{4}$ b. $6\frac{1}{4}$ c. $.6\frac{1}{4}$ d. $\frac{64}{4}$ [a] [b] [c] [d]

54. $\frac{7}{4} =$ a. 1.5   b. 1.3   c. 1.75   d. 1.7   [a] [b] [c] [d]

55. $\frac{2}{5} =$ a. .4   b. 4   c. .04   d. .25   [a] [b] [c] [d]

56. $\frac{1}{12} =$ a. $8\frac{1}{3}$ b. $.8\frac{1}{3}$ c. $.008\frac{1}{3}$ d. $.08\frac{1}{3}$ [a] [b] [c] [d]

57. $7.61 + 3\frac{2}{5} =$ a. 11.01   b. 1,101   c. 1.101   [a] [b] [c] [d]
    d. .1101

58. $44.2 \times 4\frac{1}{3} =$ a. .19139   b. 191.53   c. .019139   d. 87.7   ⓐ ⓑ ⓒ ⓓ

59. $1.2 - 1\frac{1}{10} =$ a. .11   b. .01   c. .1   d. .011   ⓐ ⓑ ⓒ ⓓ

60. $.004 =$ a. 250   b. 1.250   c. 25   d. $\frac{1}{250}$ ⓐ ⓑ ⓒ ⓓ

61. $8.003 =$ a. $8\frac{3}{1000}$ b. $\frac{83}{1000}$ c. $8\frac{3}{100}$ d. $\frac{83}{100}$ ⓐ ⓑ ⓒ ⓓ

62. $777.05 =$ a. $777\frac{5}{20}$ b. $777\frac{1}{20}$ c. 77.5   d. $777\frac{1}{200}$ ⓐ ⓑ ⓒ ⓓ

63. $.032 =$ a. $\frac{1}{31}$ b. $\frac{4}{125}$ c. $\frac{4}{1250}$ d. $\frac{2}{625}$ ⓐ ⓑ ⓒ ⓓ

64. $.33\frac{1}{3} =$ a. $\frac{1}{9}$ b. $33\frac{1}{3}$ c. $\frac{100}{3}$ d. $\frac{1}{3}$ ⓐ ⓑ ⓒ ⓓ

65. $.24\frac{1}{4} =$ a. $\frac{1}{4}$ b. $\frac{25}{4}$ c. $\frac{97}{4}$ d. $\frac{97}{400}$ ⓐ ⓑ ⓒ ⓓ

66. $.04\frac{1}{6} =$ a. $\frac{25}{6}$ b. $\frac{1}{24}$ c. $4\frac{1}{6}$ d. $.4\frac{1}{6}$ ⓐ ⓑ ⓒ ⓓ

67. $.6\frac{1}{4} =$ a. $\frac{1}{16}$ b. $\frac{25}{4}$ c. $6\frac{1}{4}$ d. $6.\frac{1}{4}$ ⓐ ⓑ ⓒ ⓓ

68. thirty-nine percent =   a. 39%   b. .39%   c. 3.9%   d. 39.00   ⓐ ⓑ ⓒ ⓓ

69. five and four-tenths percent =   a. 5.4   b. .54%   c. 5.410%   d. 5.4%   ⓐ ⓑ ⓒ ⓓ

70. one-third percent =   a. .3%   b. $\frac{1}{30}$%   c. $\frac{1}{3}$%   d. $.\frac{1}{3}$%   ⓐ ⓑ ⓒ ⓓ

71. twenty-eight and two-fifths percent =   a. $.28\frac{2}{5}$%   b. $2.8\frac{2}{5}$ c. $28\frac{2}{5}$%   d. 28.2%   ⓐ ⓑ ⓒ ⓓ

72. $300\% =$ a. 300   b. 30   c. 3.3   d. 3   ⓐ ⓑ ⓒ ⓓ

73. $17.5\% =$ a. 17.5   b. .175   c. 1.75   d. 17,500   ⓐ ⓑ ⓒ ⓓ

74. $5.85\% =$ a. .0585   b. 5.85   c. 58.5%   d. .585   ⓐ ⓑ ⓒ ⓓ

75. $80.25\% =$ a. $80\frac{1}{4}$ b. $.80\frac{1}{4}$ c. .8025   d. 8,025   ⓐ ⓑ ⓒ ⓓ

76. $5\frac{1}{2}\% =$ a. .55   b. .055   c. 5.5   d. 55%   ⓐ ⓑ ⓒ ⓓ

77. $2\frac{1}{3}\% =$ a. 2,330   b. .233   c. .0233   d. 2.33   ⓐ ⓑ ⓒ ⓓ

78. $105\frac{3}{4}\% =$ a. .10575   b. 10,575   c. .010575   d. 1.0575   ⓐ ⓑ ⓒ ⓓ

79. $\frac{1}{4}\% =$ a. $\frac{1}{4}$ b. .25   c. .0025   d. 25   ⓐ ⓑ ⓒ ⓓ

80. $.75 =$ a. 75%   b. .75%   c. .075%   d. .750%   ⓐ ⓑ ⓒ ⓓ

81. .2 =   a. 20%   b. .2%   c. 2%   d. 200% $\boxed{a}$ $\boxed{b}$ $\boxed{c}$ $\boxed{d}$

82. .208 =   a. .208%   b. 20.8%   c. 208% $\boxed{a}$ $\boxed{b}$ $\boxed{c}$ $\boxed{d}$
         d. 2.08%

83. $.66\frac{4}{5}\%$ =   a. .668%   b. 66.8%   c. 6.68% $\boxed{a}$ $\boxed{b}$ $\boxed{c}$ $\boxed{d}$
         d. 668%

84. $\frac{3}{10}$ = (percent equivalent) $\boxed{a}$ $\boxed{b}$ $\boxed{c}$ $\boxed{d}$
         a. $\frac{3}{10}\%$ b. .3%   c. 30%   d. 3%

85. $\frac{7}{50}$ =   a. .14%   b. $\frac{7}{50}\%$ c. .1400%   d. 14% $\boxed{a}$ $\boxed{b}$ $\boxed{c}$ $\boxed{d}$

86. $12\frac{1}{2}$ =   a. 12.50%   b. 12.5%   c. 1.25% $\boxed{a}$ $\boxed{b}$ $\boxed{c}$ $\boxed{d}$
         d. 1,250%

87. $3\frac{3}{7}$ =   a. $3.42\frac{6}{7}\%$ b. 342.67%   c. 342% $\boxed{a}$ $\boxed{b}$ $\boxed{c}$ $\boxed{d}$
         d. 34,267%

88. 18%   a. 9.5   b. 18   c. .18%   d. $\frac{9}{50}$ $\boxed{a}$ $\boxed{b}$ $\boxed{c}$ $\boxed{d}$

89. 375% =   a. $3\frac{7}{5}$ b. 375   c. $3\frac{3}{4}$ d. 37.5 $\boxed{a}$ $\boxed{b}$ $\boxed{c}$ $\boxed{d}$

90. 210% =   a. 210   b. $2\frac{1}{10}$ c. $\frac{210}{1000}$ d. 2.11 $\boxed{a}$ $\boxed{b}$ $\boxed{c}$ $\boxed{d}$

91. 78% =   a. $\frac{78}{1000}$ b. $\frac{78}{10}$ c. $\frac{78}{50}$ d. $\frac{39}{50}$ $\boxed{a}$ $\boxed{b}$ $\boxed{c}$ $\boxed{d}$

92. $7\frac{3}{4}\%$ =   a. 31.40   b. $\frac{31}{100}$ c. $\frac{31}{400}$ d. $\frac{31}{4000}$ $\boxed{a}$ $\boxed{b}$ $\boxed{c}$ $\boxed{d}$

93. $1\frac{2}{5}\%$ =   a. $\frac{7}{500}$ b. $\frac{7}{50}$ c. 750   d. $\frac{7}{2}$ $\boxed{a}$ $\boxed{b}$ $\boxed{c}$ $\boxed{d}$

94. $\frac{7}{10}\%$ =   a. $\frac{7}{10}$ b. $\frac{7}{1000}$ c. $\frac{7}{100}$ d. $\frac{7}{10000}$ $\boxed{a}$ $\boxed{b}$ $\boxed{c}$ $\boxed{d}$

95. $15\frac{1}{2}\%$ =   a. $\frac{31}{2}$ b. $\frac{31}{200}$ c. $\frac{15}{2}$ d. $\frac{31}{20}$ $\boxed{a}$ $\boxed{b}$ $\boxed{c}$ $\boxed{d}$

96. 97% of 400 is $\boxed{a}$ $\boxed{b}$ $\boxed{c}$
         a. equal to 400.   b. less than 400.
         c. greater than 400.

97. 105% of 36 is $\boxed{a}$ $\boxed{b}$ $\boxed{c}$
         a. equal to 36.   b. less than 36.
         c. greater than 36.

98. 6% of 110 is $\boxed{a}$ $\boxed{b}$ $\boxed{c}$
         a. equal to 110.   b. less than 110.
         c. greater than 110.

99. 253% of 3 is $\boxed{a}$ $\boxed{b}$ $\boxed{c}$
         a. equal to 3.   b. less than 3.
         c. greater than 3.

100. What is 20% of 225? $\boxed{a}$ $\boxed{b}$ $\boxed{c}$ $\boxed{d}$
         a. 45   b. 450   c. 4.5   d. .45

101. $66\frac{2}{3}\%$ of 33 = $\boxed{a}$ $\boxed{b}$ $\boxed{c}$ $\boxed{d}$
         a. 219.978   b. 21.9978   c. 2,199.78
         d. 21,997.8

129

102. $\frac{1}{2}$% of 100 =

     a. .5   b. .05   c. .005   d. .0005

a b c d

103. 8 is 10% of _____.

     a. 8   b. .8   c. 80   d. .08

a b c d

104. 9 is 15% of what number?

     a. .6   b. .06   c. 60   d. 6

a b c d

105. 306 is 90% of _____.

     a. 34   b. 3.4   c. .34   d. 340

a b c d

106. 3.1 is 20% of what number?

     a. 15.5   b. 1.55   c. 155   d. 1.55

a b c d

107. 2 is _____% of 10.

     a. 2%   b. 20%   c. 2,000%

     d. 200%

a b c d

108. What percent of 126 is 63?

     a. 5%   b. .5%   c. .05%   d. 50%

a b c d

109. 1.8 is _____% of 720.

     a. 25%   b. 2.5%   c. $\frac{1}{4}$%   d. $\frac{1}{40}$%

a b c d

110. 75 is what percent of 60?

     a. 12.5%   b. 75%   c. 7.5%

     d. 125%

a b c d

Solve these word problems.

111. The state employment commission reported that there were 12,800 eligible unemployed persons in the county during March. If this was 3.2 percent of the work force, how many *employed* persons were there during the period?

     a. 40,960   b. 400,000   c. 440,960   d. 387,200

a b c d

112. The sales tax on the Greenes' new car was $1,158.75. If sales tax in their state is 5 percent, what was the price of their car before the tax was added?

     a. $23,175   b. $12,166   c. $5,793.75

     d. $57.93

a b c d

113. What was the price of the Greenes' car including tax?

     a. $24,333.75   b. $13,324.75   c. $5,952.50

     d. $1,216.68

a b c d

114. An aerosol can of flea spray contains active ingredients and inert (inactive) ingredients. Active ingredients make up only 1.65% of the six ounces of spray. What is the weight of the inert ingredients?
a. 9.9 ounces   b. 14.3 ounces   c. .99 ounces
d. 5.901 ounces

a b c d

115. Than Phan, a bookkeeper, makes $7.15 an hour. How much does he make for working a 40-hour week?
a. $286   b. $16   c. $43.85   d. $2,860

a b c d

116. Ms. Sanchez lives 16.4 miles away from Baker Clinic, where she goes for physical therapy twice a week. How far does she travel to and from the clinic each week?
a. 32.8 miles   b. 65.6 miles   c. 131.2 miles
d. 8.2 miles

a b c d

117. Taxi fare is $1.00 for the first tenth of a mile and $.10 for each additional tenth of a mile. Cathy Shioji took a taxi from her house to the airport 5.3 miles away. What was the cost of the trip?
a. $6.20   b. $10.07   c. $5.30   d. $6.30

a b c d

118. Ms. Shioji gave the taxi driver a tip of $.80. How much money did she have left from the $10 bill she used to pay the driver?
a. $3.00   b. $.07   c. $3.90   d. $1.80

a b c d

119. In one school district, teachers with no teaching experience are paid $16,998 a year. How much will this be per month on a 12-month pay basis?
a. $1,699   b. $1,699.80   c. $1,416.50   d. $1,416

a b c d

120. Sacha's car holds 24 gallons of gasoline. If he can drive 420 miles on that amount, how many miles per gallon does his car get?
a. 5.714   b. 1.75   c. 57.14   d. 17.5

a b c d

121. Which is the better buy: a 16-ounce can of stewed tomatoes costing $.58 or a 24-ounce can costing $.89?
a. the 16-ounce can   b. the 24-ounce can

a b

122. Joe Ryan bought $1\frac{1}{2}$ pounds of pistachios at $4.90 a pound. How much did he pay for the nuts?
a. $14.70   b. $7.35   c. $4.90   d. $9.80

a b c d

123. Gregorio needed 2 pounds of ground beef and one pound of pork sausage for his meat loaf. Ground beef was on sale for $2.09 a pound and the sausage was $2.39 a pound. How much did the meat for the meal cost?
a. $4.48   b. $6.87   c. $4.78   d. 6.57

a b c d

124. Anadine's hourly pay is $8.20 an hour. How much would she receive for working $26\frac{1}{4}$ hours?
    a. $34.45   b. $21.32   c. $215.25   d. $213.20

    a b c d

125. Anadine is paid $1\frac{1}{2}$ times her regular hourly rate for time worked past 40 hours in a week. How much would she be paid for working 45 hours?
    a. $369   b. $553.50   c. $448.30   d. $389.50

    a b c d

126. Work shirts are on sale at store A for 30% off. Store B is selling the same shirts for $\frac{1}{3}$ off. If both stores had the same original price, where could you get the better deal?
    a. store A   b. store B

    a b

127. Jane Freeman receives a commission of 3% on her sales. Last month her sales totaled $12,863. What is the amount of her commission?
    a. $38.79   b. $385.89   c. $385.08   d. $3,858.90

    a b c d

128. Adrian has 4% of his monthly pay deducted for savings. The amount is $100 per month. What is Adrian's annual salary?
    a. $16,000   b. $15,000   c. $26,000   d. $30,000

    a b c d

129. On August 1 Tracy's bank balance was $2,844.68. He wrote a check for $126 for a coat. What was his new balance?
    a. $2,718.68   b. $271.86   c. $2,970.68
    d. $297.06

    a b c d

130. Beatrice's salary is $1,416.75 a month. In three weeks, she will receive a raise of $75 per month. What will her annual salary be?
    a. $17,001   b. $1,491.75   c. $17,076
    d. $17,901

    a b c d

131. At the pet store Jayson spent $3.66 for a leash, $2.99 for a collar, $1.19 for a toy, $24.95 for a dog bed, and $2.09 for pet shampoo. He also bought four cartons of dog food costing $.89 each. Jayson paid for his purchases with a $50 bill. How much change did he receive?
    a. $11.56   b. $38.44   c. $35.77   d. $14.23

    a b c d

132. There are 10 wieners in a pound of all-beef wieners costing $2.29. What is the cost of 2 wieners?
    a. 22.9¢   b. 45.8¢   c. 35¢   d. 91.6¢

    a b c d

133. Carlo Antorelli works on commission. Last month he sold $24,000 worth of computer software. If he made $1,920, what percent commission did he receive?
    a. 12½%   b. 10%   c. 8%   d. 8¼%

    a b c d

134. Bus fare from Smalltown to Midtown has increased from $4.80 to $6.00. What is the percent of increase?
a. 1.25%   b. 20%   c. 25%   d. 8%

`a` `b` `c` `d`

135. Kevin scored 85% on his American history test. He answered 34 questions correctly. How many questions were on the test?
a. 51   b. 40   c. 78   d. 25

`a` `b` `c` `d`

136. Chris plays intramural basketball. So far this season she has made 80 field goals. This is 40% of the number of field goals she has attempted. How many field goals has she attempted this season?
a. 32   b. 50   c. 200   d. 140

`a` `b` `c` `d`

137. A store advertised tires at 20% off. What was the original price of a tire that was marked $15 off?
a. $85   b. $75   c. $80   d. $90

`a` `b` `c` `d`

138. A theater seats 320 people. If the theater is filled to 75% of capacity, how many seats are empty?
a. 240   b. 320   c. 395   d. 80

`a` `b` `c` `d`

139. Ana Cook received a commission of $4,200 for selling a $60,000 house. At that rate, what would be her commission for selling a $100,000 house?
a. $7,000   b. $6,000   c. $10,000   d. $9,000

`a` `b` `c` `d`

140. Johanne paid $10,500 for a car. Its value has now decreased by $3,150. By what percent has the car depreciated in value?
a. 15%   b. 30%   c. 70%   d. 33⅓%

`a` `b` `c` `d`

141. In a sample of ore weighing 98 kilograms, there was 4.5% metal. How many kilograms of metal were in the sample?
a. 2.177   b. 21.77   c. 102.5   d. 4.41

`a` `b` `c` `d`

142. The price of a scarf was marked down $1.20. The original price of the scarf was $6. What was the amount of discount in percent?
a. 25%   b. 80%   c. 75%   d. 20%

`a` `b` `c` `d`

143. Marlo works 40 hours a week. Each day she makes $40.56. This is 20% of a week's pay. How much does Marlo make each week?
a. $324.48   b. $162.24   c. $203.25   d. $202.80

`a` `b` `c` `d`

144. If your monthly salary of $1,575 is increased by 7%, how much will you make each month?
a. $1,685.25   b. $110.25   c. $1,879.99
d. $1,576.07

`a` `b` `c` `d`

145. A 5-pound roast costs $12.95. What is the price per pound?

a. $80.94   b. $2.19   c. $2.79   d. $2.59   ⬚a ⬚b ⬚c ⬚d

146. When Greg bought his car, the odometer showed 56,004.6 miles. At the end of one month it read 57,242.8 miles. How far did Greg drive his car that month?

a. 1,238.2 miles   b. 123.82 miles   c. 1.2382 miles

d. 12.382 miles   ⬚a ⬚b ⬚c ⬚d

Page 1. 1. $\frac{8}{100}$; 2. $\frac{11}{100}$; 3. $\frac{25}{1000}$; 4. $\frac{983}{1000}$; 5. $\frac{2402}{10000}$; 6. $\frac{7153}{10000}$; 7. $\frac{1}{100000}$; 8. $\frac{56789}{100000}$; 9. .0071, $\frac{71}{10000}$; 10. .02004, $\frac{2004}{100000}$; 11. .8, $\frac{8}{10}$; 12. .005, $\frac{5}{1000}$; 13. a, b; 14. b, c; 15. Answer given in text. 16. $\frac{5}{100}$; 17. .17, seventeen hundredths; 18. $\frac{57}{1000}$, fifty-seven thousandths; 19. .00379, $\frac{379}{100000}$

Page 2. A. 1. $\frac{3}{10}$; 2. $\frac{4}{10}$; 3. $\frac{8}{10}$; 4. $\frac{7}{10}$; 5. $\frac{1}{10}$; 6. $\frac{6}{10}$; 7. $\frac{2}{10}$; 8. $\frac{1}{10}$; 9. $\frac{9}{10}$; 10. $\frac{5}{10}$; 11. $\frac{4}{10}$; 12. $\frac{8}{10}$; 13. $\frac{5}{10}$; 14. $\frac{3}{10}$; 15. $\frac{7}{10}$; 16. $\frac{2}{10}$; B. 1. $\frac{7}{10}$; 2. $\frac{3}{10}$; 3. $\frac{2}{10}$; 4. $\frac{1}{10}$; 5. $\frac{5}{10}$; 6. $\frac{9}{10}$; 7. $\frac{6}{10}$; 8. $\frac{4}{10}$; 9. $\frac{8}{10}$; 10. $\frac{2}{10}$

Page 3. A. 1. $\frac{25}{100}$; 2. $\frac{16}{100}$; 3. $\frac{1}{100}$; 4. $\frac{8}{100}$; 5. $\frac{60}{100}$; 6. $\frac{20}{100}$; 7. $\frac{9}{100}$; 8. $\frac{2}{100}$; 9. $\frac{33}{100}$; 10. $\frac{85}{100}$; 11. $\frac{7}{100}$; 12. $\frac{50}{100}$; 13. $\frac{39}{100}$; 14. $\frac{97}{100}$; 15. $\frac{11}{100}$; 16. $\frac{75}{100}$; 17. $\frac{44}{100}$; 18. $\frac{51}{100}$; 19. $\frac{23}{100}$; 20. $\frac{70}{100}$; B. 1. $\frac{34}{100}$; 2. $\frac{71}{100}$; 3. $\frac{6}{100}$; 4. $\frac{56}{100}$; 5. $\frac{33}{100}$; 6. $\frac{40}{100}$; 7. $\frac{75}{100}$; 8. $\frac{52}{100}$; 9. $\frac{16}{100}$; 10. $\frac{8}{100}$; 11. $\frac{98}{100}$; 12. $\frac{45}{100}$

Page 4. A. 1. $\frac{6}{1000}$; 2. $\frac{8}{1000}$; 3. $\frac{238}{1000}$; 4. $\frac{562}{1000}$; 5. $\frac{600}{1000}$; 6. $\frac{200}{1000}$; 7. $\frac{25}{1000}$; 8. $\frac{40}{1000}$; 9. $\frac{70}{1000}$; 10. $\frac{90}{1000}$; 11. $\frac{186}{1000}$; 12. $\frac{75}{1000}$; 13. $\frac{217}{1000}$; 14. $\frac{50}{1000}$; 15. $\frac{101}{1000}$; 16. $\frac{868}{1000}$; 17. $\frac{225}{1000}$; 18. $\frac{458}{1000}$; B. 1. $\frac{101}{1000}$; 2. $\frac{30}{1000}$; 3. $\frac{683}{1000}$; 4. $\frac{135}{1000}$; 5. $\frac{4}{1000}$; 6. $\frac{602}{1000}$; 7. $\frac{500}{1000}$; 8. $\frac{984}{1000}$; 9. $\frac{78}{1000}$; 10. $\frac{2}{1000}$; 11. $\frac{410}{1000}$; 12. $\frac{930}{1000}$; 13. $\frac{705}{1000}$; 14. $\frac{706}{1000}$

Page 5. A. 1. $\frac{8}{10000}$; 2. $\frac{30}{10000}$; 3. $\frac{33}{10000}$; 4. $\frac{2}{10000}$; 5. $\frac{445}{10000}$; 6. $\frac{635}{10000}$; 7. $\frac{2643}{10000}$; 8. $\frac{3030}{10000}$; 9. $\frac{5002}{10000}$; 10. $\frac{4006}{10000}$; 11. $\frac{234}{10000}$; 12. $\frac{5601}{10000}$; 13. $\frac{5410}{10000}$; 14. $\frac{7154}{10000}$; B. 1. $\frac{2304}{10000}$; 2. $\frac{3333}{10000}$; 3. $\frac{613}{10000}$; 4. $\frac{6011}{10000}$; 5. $\frac{1}{10000}$; 6. $\frac{4432}{10000}$; 7. $\frac{5306}{10000}$; 8. $\frac{106}{10000}$; 9. $\frac{103}{10000}$; 10. $\frac{10}{10000}$; 11. $\frac{1011}{10000}$; 12. $\frac{5050}{10000}$; 13. $\frac{2410}{10000}$; 14. $\frac{6162}{10000}$

Page 6. A. 1. $\frac{44}{100000}$; 2. $\frac{18}{100000}$; 3. $\frac{40003}{100000}$; 4. $\frac{60002}{100000}$; 5. $\frac{200}{100000}$; 6. $\frac{63000}{100000}$; 7. $\frac{1023}{100000}$; 8. $\frac{408}{100000}$; 9. $\frac{2}{100000}$; 10. $\frac{9}{100000}$; 11. $\frac{4014}{100000}$; 12. $\frac{142}{100000}$; 13. $\frac{72710}{100000}$; 14. $\frac{36891}{100000}$; 15. $\frac{80345}{100000}$; 16. $\frac{5}{100000}$; B. 1. $\frac{52301}{100000}$; 2. $\frac{100}{100000}$; 3. $\frac{6304}{100000}$; 4. $\frac{9}{100000}$; 5. $\frac{32789}{100000}$; 6. $\frac{10019}{100000}$; 7. $\frac{1}{100000}$; 8. $\frac{21006}{100000}$; 9. $\frac{10006}{100000}$; 10. $\frac{98765}{100000}$; 11. $\frac{1023}{100000}$; 12. $\frac{12345}{100000}$; 13. $\frac{65423}{100000}$; 14. $\frac{242}{100000}$

Page 7. A. 1. .0008, $\frac{8}{10000}$; 2. .023, $\frac{23}{1000}$; 3. .4, $\frac{4}{10}$; 4. .02038, $\frac{2038}{100000}$; 5. .55, $\frac{55}{100}$; 6. .062, $\frac{62}{1000}$; 7. .8324, $\frac{8324}{10000}$; B. 1. .0068, $\frac{68}{10000}$; 2. .5, $\frac{5}{10}$; 3. .056, $\frac{56}{1000}$; 4. .12038, $\frac{12038}{100000}$; 5. .004, $\frac{4}{1000}$; 6. .01036, $\frac{1036}{100000}$; 7. .0433, $\frac{433}{10000}$

Page 8. A. 1. 4.23, $4\frac{23}{100}$; 2. 163.008, $163\frac{8}{1000}$; 3. 5,008.2463, $5,008\frac{2463}{10000}$; 4. 44.00004, $44\frac{4}{100000}$; B. 1. 268.04, $268\frac{4}{100}$; 2. 10.1002, $10\frac{1002}{10000}$; 3. 94.076, $94\frac{76}{1000}$; 4. 500.2, $500\frac{2}{10}$

Page 9. A. 2. $\frac{4}{10}$; 3. .05, 5 hundredths; 4. $\frac{23}{1000}$, 23 thousandths; 5. .00426, $\frac{426}{100000}$; 6. $\frac{24}{100000}$, 24 hundred-thousandths; 7. .01045, 1,045 hundred-thousandths; 8. .0530, $\frac{530}{10000}$; 9. $\frac{4}{1000}$, 4 thousandths; 10. .802, 802 thousandths; 11. .016, $\frac{16}{1000}$; 12. $\frac{141}{1000}$, one hundred forty-one thousandths; 13. .84, eighty-four hundredths; 14. .00987, $\frac{987}{100000}$; 15. .7, seven tenths; 16. $\frac{5}{10}$, five tenths; 17. .21, $\frac{21}{100}$; 18. .666, six hundred sixty-six thousandths; 19. $\frac{19}{1000}$, nineteen thousandths; 20. .0702, $\frac{702}{10000}$; 21. .091, $\frac{91}{1000}$

Page 10. B. 2. .7, 7 tenths; 3. .04, $\frac{4}{100}$; 4. $\frac{6}{10000}$, 6 ten-thousandths; 5. 16.002, 16 and 2 thousandths; 6. 12.03, $12\frac{3}{100}$; 7. 436.03, $436\frac{3}{100}$, 436 and 3 hundredths; 8. 213.0025, 213 and 25 ten-thousandths; 9. 23.023, $23\frac{23}{1000}$; 10. 1,833$\frac{18}{1000}$, 1,833 and 18 thousandths; 11. .49, forty-nine hundredths; 12. 77.6, $77\frac{6}{10}$; 13. 240$\frac{13}{1000}$, 240 and 13 thousandths; 14. 66.433, 66 and 433 thousandths; 15. 20.0011, 20$\frac{11}{10000}$; 16. $\frac{76}{10000}$, 76 ten-thousandths; 17. 39.62, 39 and 62 hundredths; 18. 18.16, $18\frac{16}{100}$; 19. $\frac{45}{100}$, forty-five hundredths; 20. 2,465.9870, 2,465 and 9870 ten-thousandths

Page 11. Answers will vary.

Page 12. 1. a, c; 2. b, d; 3. true; 4. true; 5. true; 6. true; 7. equal; 8. not equal; 9. not equal; 10. equal; 11. .9; 12. .4; 13. .149; 14. $\frac{10}{100}$, $\frac{1}{10}$; 15. $4\frac{8}{100}$, $4\frac{2}{25}$; 16. .127; 17. 16.7

Page 13. 2. $6.03, $6\frac{3}{100}$; 3. $20.40, $20\frac{40}{100}$; 4. $108.25, $108\frac{25}{100}$; 5. $4.05, $4\frac{5}{100}$; 6. $1.99, $1\frac{99}{100}$; 7. $7.17, $7\frac{17}{100}$

Page 14. A. 1. true; 2. true; 3. true; 4. true; 5. true; 6. true; 7. true; 8. true; 9. true. B. 1. true; 2. true; 3. true; 4. true; 5. true; 6. true; 7. true; 8. true

Page 15. A. 1. .8 = $\frac{8}{10}$ = $\frac{4}{5}$, .80 = $\frac{80}{100}$ = $\frac{4}{5}$, equal; 2. .25 = $\frac{25}{100}$ = $\frac{1}{4}$, .025 = $\frac{250}{1000}$ = $\frac{1}{4}$, equal; 3. .02 = $\frac{2}{100}$ = $\frac{1}{50}$, .20 = $\frac{20}{100}$ = $\frac{1}{5}$, not equal. B. 1. .6 = $\frac{6}{10}$ = $\frac{3}{5}$, .60 = $\frac{60}{100}$ = $\frac{3}{5}$, equal; 2. .08 = $\frac{8}{100}$ = $\frac{2}{25}$, .80 = $\frac{80}{100}$ = $\frac{4}{5}$, not equal; 3. .002 = $\frac{2}{1000}$ = $\frac{1}{500}$, .0020 = $\frac{20}{10000}$ = $\frac{1}{500}$, equal

Page 16. A. The following decimals should be circled: 1. .65; 2. .9; 3. .12; 4. .53; 5. .77; 6. .54;

7. .80; **8.** .13. **B.** The following decimals should be circled: **1.** .5; **2.** .2; **3.** .32; **4.** .6; **5.** .1

Page 17. **A.** **2.** 25, 25., 25.00; **3.** 633, 633.0, 633.00; **4.** 400., 400.0, 400.00; **5.** 1, 1., 1.0; **6.** 75., 75.0, 75.00; **7.** 36, 36.0, 36.00. **B.** **1.** 44, 44., 44.00; **2.** 750., 750.0, 750.00; **3.** 2, 2.0, 2.00; **4.** 13, 13., 13.0; **5.** 10., 10.0, 10.00; **6.** 234., 234.0, 234.00

Page 18. **A.** **2.** $\frac{5}{10}$, $\frac{1}{2}$; **3.** $20\frac{4}{100}$, $20\frac{1}{25}$; **4.** $1\frac{6}{10}$, $1\frac{3}{5}$; **5.** $18\frac{2}{10}$, $18\frac{1}{5}$; **6.** $123\frac{25}{1000}$, $123\frac{1}{40}$; **7.** $4\frac{25}{100}$, $4\frac{1}{4}$. **B.** **1.** $\frac{15}{100}$, $\frac{3}{20}$; **2.** $53\frac{80}{100}$, $53\frac{4}{5}$; **3.** $\frac{40}{100}$, $\frac{2}{5}$; **4.** $4\frac{2}{10}$, $4\frac{1}{5}$; **5.** $66\frac{666}{1000}$, $66\frac{333}{500}$; **6.** $74\frac{82}{100}$, $74\frac{41}{50}$

Page 19. **A.** **1.** true; **2.** true; **3.** true; **4.** false; **5.** true; **6.** true; **7.** false; **8.** true; **9.** true; **10.** false. **B.** **1.** .43; **2.** .08; **3.** 14; **4.** 2.6; **5.** 3.18; **6.** equal; **7.** equal; **8.** .9; **9.** .45; **10.** .2; **11.** .75

Page 20. **C.** **1.** equal; **2.** 14; **3.** equal; **4.** .8; **5.** equal; **6.** equal; **7.** equal; **8.** equal; **9.** $18\frac{4}{10}$; **10.** equal; **11.** equal; **12.** .3; **13.** 18; **14.** equal; **15.** .060; **16.** .756; **17.** equal; **18.** equal; **19.** equal; **20.** equal; **21.** 66.2; **22.** .77; **23.** 16.1; **24.** equal; **25.** $\frac{4}{5}$; **26.** 202.2

Page 21. No answers are required.

Page 22. **1.** 309.887; **2.** 549.03; **3.** 14,175.10; **4.** 1,043.963; **5.** 33.341; **6.** 5,046.165; **7.** 21.7022; **8.** 414.3843; **9.** 28.411; **10.** $24.87; **11.** $95.66; **12.** 163.203; **13.** b; **14.** d; **15.** b

Page 23. **A.** **1.** 176.186; **2.** .859; **3.** $6.75; **4.** $25.85; **5.** $129.88; **6.** $30.75; **7.** 4.1747; **8.** $600.10; **9.** 4.6432. **B.** **1.** 398.50; **2.** .459; **3.** $8.66; **4.** $54.93; **5.** $129.19; **6.** 1,078.149

Page 24. **A.** **1.** 1,408.548; **2.** 19.465; **3.** $124.70; **4.** $1.04. **B.** **1.** 122.14; **2.** 23.056; **3.** $4.19; **4.** $1.42

Page 25. **A.** **1.** 64.798; **2.** 358.706; **3.** $9.94; **4.** $13.84. **B.** **1.** 473.063; **2.** 59.048; **3.** $72.37; **4.** $55.16; **5.** 300.203; **6.** $133.43

Page 26. **1.** $807.93; **2.** 96.5 miles; **3.** $375.46; **4.** $1,458.34; **5.** 10 miles; **6.** $244.81; **7.** $9

Page 27. **A.** **1.** 142.673; **2.** 140.006; **3.** 8.7829; **4.** 85.021; **5.** .271; **6.** 7.961; **7.** 215.2864; **8.** 56.8578; **9.** 303.1037; **10.** 807.3911.

Page 28. **B.** **1.** .53; **2.** 370.141; **3.** $23.63; **4.** $180.35; **5.** $24.48; **6.** 143.803; **7.** 131.103; **8.**

53.21163; **9.** 142.66; **10.** 19.322; **11.** 8.013; **12.** 1,043.366; **13.** 100.86603; **14.** 5,146.063

Page 29. **1.** $565.70; **2.** $156.03

Page 30. **1.** .8794; **2.** $288.89; **3.** $1.04; **4.** 155.624; **5.** $794.15; **6.** 407.160; **7.** $2.99; **8.** .0084; **9.** 51.12; **10.** $5.05; **11.** 1.9316; **12.** $9.37; **13.** 637.941; **14.** $1.15; **15.** $8.06; **16.** 68.6939; **17.** b

Page 31. **A.** **1.** 11.221; **2.** 3.0733; **3.** $52.99; **4.** $94.40; **5.** $5.25; **6.** 24.706. **B.** **1.** 1,240.5; **2.** 2.1014; **3.** $126.17; **4.** $22.93; **5.** $4.28; **6.** 18.247

Page 32. **A.** **1.** 18.054; **2.** 39.773; **3.** $562.04; **4.** $9.05. **B.** **1.** 9.503; **2.** 8.15; **3.** $422.32; **4.** $15.49

Page 33. **A.** **1.** 37.832; **2.** 2.738; **3.** $.65; **4.** $.50; **5.** $11.57; **6.** 20.936. **B.** **1.** 33.826; **2.** 8.3567; **3.** $1.33; **4.** $3.26

Page 34. **1.** .182 inch; **2.** 4.8 miles per gallon; **3.** $3.15; **4.** $2.42; **5.** $152.53; **6.** 5 feet; **7.** $.20; **8.** $244.95; **9.** $84.86

Page 35. **1.** the catalog purchase; **2.** $16.75; **3.** $254.78; **4.** .16; **5.** .17

Page 36. **A.** **1.** 122.182; **2.** 356.413; **3.** $15.72; **4.** $3.56; **5.** $257.06; **6.** 155.94; **7.** 82.7309; **8.** $52.42; **9.** $86.25; **10.** 451.291; **11.** 21.984; **12.** $44.64; **13.** 536.06; **14.** 58.045; **15.** 33.1; **16.** 15.811; **17.** $72.78; **18.** 535.054; **19.** 216.916; **20.** 73.8845; **21.** $28.49; **22.** $29.36; **23.** $45.35; **24.** $4.15.

Page 37. **B.** **1.** 5.232; **2.** 2,612.56; **3.** $46.19; **4.** $7.31; **5.** $120.54; **6.** 27.9; **7.** $82.45; **8.** $14.01; **9.** 838.824; **10.** 197.007; **11.** $2.49; **12.** 168.35; **13.** 9.837; **14.** 33.9931; **15.** 5.14; **16.** 22.976; **17.** $647.44; **18.** 105.655; **19.** 9.1483; **20.** $.08; **21.** 48.46; **22.** 65.59; **23.** 575.0696; **24.** $18.84

Page 38. **1.** $161.87; **2.** $49.65; **3.** $16.88; **4.** $380.72; **5.** $498.80; **6.** $2,052.91

Page 39. **1.** 4; **2.** 0; **3.** 1; **4.** 5; **5.** 2,184; **6.** 27.7776; **7.** 9.14305; **8.** 4.59464; **9.** .05; **10.** .09; **11.** .0105; **12.** .03672; **13.** .19236; **14.** .895114; **15.** 11.984; **16.** .000152; **17.** a; **18.** b

Page 40. **A.** **1.** 1; **2.** 2; **3.** 4; **4.** 0; **5.** 5; **6.** 1; **7.** 1; **8.** 5; **9.** 4; **10.** 0; **11.** 3; **12.** 3; **13.** 2; **14.** 3; **15.** 0; **16.** 2; **17.** 2; **18.** 2; **19.** 3; **20.** 3; **21.** 3; **22.** 4; **23.** 4; **24.** 4. **B.** **1.** 3; **2.** 1; **3.** 4; **4.** 2;

5. 0; **6.** 4; **7.** 2; **8.** 1; **9.** 1; **10.** 2; **11.** 2; **12.** 2; **13.** 2; **14.** 0; **15.** 1

Page 41. A. 1. 17.082; **2.** 1,514.688; **3.** 3.0912; **4.** 864.9; **5.** .198; **6.** .8064. **B. 1.** 20.864; **2.** .26487; **3.** 8.82126; **4.** 4.5864; **5.** .00289; **6.** 28,991.592

Page 42. A. 1. .08; **2.** .064; **3.** .0096; **4.** .0775; **5.** .09; **6.** .00078; **7.** .06919; **8.** .0004929; **9.** .0621. **B. 1.** .09; **2.** .099; **3.** .0135; **4.** .0832; **5.** .06; **6.** .0462

Page 43. 1. $34.40; **2.** $172; **3.** $8.25; **4.** $390; **5.** $27.50; **6.** 83 miles; **7.** $10,326.72; **8.** $60.50; **9.** $1.60; **10.** $7.14 (7.143 to nearest hundredth)

Page 44. A. 1. .0500; **2.** 60.50; **3.** 14.7492; **4.** 8.820; **5.** .03423; **6.** 8.9356; **7.** .000336; **8.** 12.068; **9.** .00346; **10.** 3.36; **11.** .07383; **12.** .82492; **13.** 57.40; **14.** 32.294; **15.** 109.2; **16.** 48.96; **17.** .0024; **18.** 11.52.

Page 45. B. 1. 32.280; **2.** .14454; **3.** .45090; **4.** 77.76; **5.** .06; **6.** .0516; **7.** 36.42; **8.** .23; **9.** .315; **10.** 24.18; **11.** 5.379; **12.** .02020; **13.** .1403; **14.** 60; **15.** 57.6; **16.** 408.096; **17.** 1.699426; **18.** 7,470.304

Page 46. 1. $330; **2.** $385; **3.** $269; **4.** $984; **5.** $\frac{1}{20}$; **6.** $81.34; **7.** $39.85

Page 47. 1. 4.4; **2.** 2.1; **3.** 2.208; **4.** .33; **5.** .5; **6.** .36; **7.** .43; **8.** .865; **9.** .3; **10.** 6; **11.** .8; **12.** 203.1; **13.** 1,010; **14.** 130; **15.** 82,000; **16.** 1,100;

Page 48. 17. 7.21; **18.** 7.5; **19.** 145.46; **20.** .17; **21.** a; **22.** c; **23.** d; **24.** c; **25.** b; **26.** a; **27.** b; **28.** b

Page 49. A. 1. 2.2; **2.** .09; **3.** 1.1; **4.** .3; **5.** .007; **6.** .32; **B. 1.** 4.1; **2.** .21; **3.** 31.4

Page 50. A. 1. .25; **2.** 1.8; **3.** 1.2; **4.** .4; **5.** .875; **6.** .275; **7.** .2; **8.** .5; **9.** .003;

Page 51. 10. .153; **11.** .9999; **12.** .834; **13.** .145; **14.** .9; **15.** .81; **16.** .7065; **17.** .6; **18.** .303; **B. 1.** .625; **2.** 20.5; **3.** .25; **4.** .36; **5.** .8; **6.** .04; **7.** .94; **8.** .5; **9.** .5

Page 52. 1. 3; **2.** 4; **3.** 9.79; **4.** .3; **5.** 5.2; **6.** .12; **B. 1.** 33; **2.** 3; **3.** 369; **4.** .4

Page 53. A. 1. 300; **2.** 20; **3.** 200; **4.** 35; **5.** 3,000; **6.** 500; **B. 1.** 30; **2.** 300; **3.** 4,000

Page 55. A. 1. .57; **2.** .31; **3.** 18.28; **4.** .32; **5.**

32.36; **6.** 4.08; **B. 1.** 20.3; **2.** 4.2; **3.** 1.2; **4.** 13.4; **5.** 16.8; **6.** .1; **7.** 1.1; **8.** 1; **9.** .7

Page 56. $1,050.55; **2.** $3.25; **3.** 14 miles per gallon;

Page 57. 4. $.29; **5.** 5 months; **6.** $15.75; **7.** $.95; **8.** $.98; **9.** 4 pillows; **10.** $7.85; **11.** $.53; **12.** $8.75

Page 58. 1. $646.30; **2.** $197.75; **3.** $6.93; **4.** the 3-pound can; **5.** 9,550 miles

Page 60. A. 1. 3.2; **2.** .28; **3.** .5; **4.** .4; **5.** 1.2; **6.** 21; **7.** 300; **8.** .3; **9.** 1.5; **B. 1.** 1.3; **2.** .13; **3.** .5; **4.** .25; **5.** .02; **6.** 4

Page 61. C. 1. .24; **2.** 1.6; **3.** 5; **4.** 200; **5.** 39,700; **6.** .75; **7.** 20; **8.** .036; **9.** 300; **10.** .25; **11.** 10,100; **12.** 324

Page 63. 1. a. 14.3¢ per ounce, b. 13.9¢ per ounce, better buy—b; **2.** a. 10.5¢ per ounce, b. 10.32¢ per ounce, better buy—b; **3.** a. 9.83¢ per ounce, b. 11.13¢ per ounce, better buy—a; **4.** a. 4.1¢ per ounce, b. 3.06¢ per ounce, better buy—b; **5.** a. $1.20 per pound, b. $2.99 per pound better buy—a. (This may be misleading since the picnic ham contains bone and gristle. It is difficult to compare unit prices for two items when they are not alike.); **6.** a. 13.3¢ per ounce, b. 13.7¢ per ounce, better buy—a; **7.** a. 13.3¢ per ounce, b. 11.13¢ per ounce, better buy—b; **8.** a. 16.33¢ per ounce, b. 7.7¢ per ounce, better buy—b; **9.** a. 6.63¢ per ounce, b. 6.15¢ per ounce, better buy—b; **10.** a. 6.63¢ per ounce, b. 5.92¢ per ounce, better buy—b.

Page 64. 1. .6; **2.** .5833; **3.** 1.25; **4.** .5625; **5.** .14$\frac{2}{7}$; **6.** .11$\frac{1}{9}$; **7.** .83$\frac{1}{3}$; **8.** .66$\frac{2}{3}$; **9.** 14.25; **10.** 808.8; **11.** 8.59; **12.** 19.15; **13.** 90.4; **14.** 18.524; **15.** $\frac{17}{50}$; **16.** $\frac{1}{200}$; **17.** 666$\frac{1}{50}$; **18.** 9$\frac{7}{1000}$; **19.** $\frac{2}{3}$; **20.** $\frac{1}{24}$; **21.** $\frac{91}{400}$; **22.** $\frac{27}{500}$; **23.** b; **24.** c

Page 66. A. 1. .4; **2.** .0625; **3.** .7; **4.** 1.25; **5.** .875; **6.** .9375; **7.** .75; **8.** .4375; **B. 1.** .8; **2.** .1875; **3.** .9; **4.** 2.5; **5.** .625; **6.** .8125; **7.** 1.1; **8.** 1.4

Page 67. A. 1. .33$\frac{1}{3}$; **2.** .66$\frac{2}{3}$; **3.** .16$\frac{2}{3}$; **4.** .27$\frac{7}{9}$; **5.** .11$\frac{1}{9}$; **6.** .83$\frac{1}{3}$;

Page 68. 7. .14$\frac{2}{7}$; **8.** .37$\frac{1}{2}$; **9.** .28$\frac{4}{7}$; **10.** .22$\frac{2}{9}$ **B. 1.** .38$\frac{8}{9}$; **2.** .41$\frac{2}{3}$; **3.** .83$\frac{1}{3}$; **4.** .58$\frac{1}{3}$; **5.** .57$\frac{17}{19}$; **6.** .71$\frac{3}{7}$; **7.** .85$\frac{5}{7}$; **8.** .88$\frac{8}{9}$; **9.** .87$\frac{1}{2}$; **10.** .90$\frac{10}{11}$

Page 69. A. 1. 5.75; **2.** 25.62; **3.** 13.33$\frac{1}{3}$; **4.**

$3.11\frac{1}{9}$; **5.** 148.6; **6.** 17.25; **7.** $10.37\frac{1}{2}$; **8.** $7.46\frac{2}{3}$;

Page 70. **9.** 20.7; **10.** $64.77\frac{7}{9}$. **B.** **1.** 7.2; **2.** 14.375; **3.** $52.66\frac{2}{3}$; **4.** $7.83\frac{1}{3}$; **5.** 216.75; **6.** 101.125; **7.** $61.66\frac{2}{3}$; **8.** 17.4; **9.** $48.33\frac{1}{3}$; **10.** $87.16\frac{2}{3}$

Page 71. **A.** **1.** 32.5; **2.** 3.59; **3.** 11.13; **4.** 8; **5.** 11.352; **6.** 6.4; **7.** 10.26; **8.** 61.875; **9.** 85.86; **10.** 31.95;

Page 72. **11.** 102.08; **12.** 198; **B.** **1.** 34.5; **2.** 8.12; **3.** 92.02; **4.** 3.025; **5.** 20.71; **6.** .44; **7.** 12.32; **8.** .85656; **9.** 62.8; **10.** .077; **11.** .07; **12.** 6.9

Page 73. **A.** **1.** $\frac{1}{10}$; **2.** $\frac{38}{100} = \frac{19}{50}$; **3.** $\frac{4}{10} = \frac{2}{5}$; **4.** $\frac{25}{1000} = \frac{1}{40}$; **5.** $\frac{25}{100} = \frac{1}{4}$; **6.** $338\frac{4}{100} = 338\frac{1}{25}$; **7.** $\frac{15}{100} = \frac{3}{20}$; **8.** $\frac{5}{100} = \frac{1}{20}$; **9.** $\frac{6}{10} = \frac{3}{5}$; **10.** $\frac{75}{100} = \frac{3}{4}$; **11.** $\frac{32}{100} = \frac{8}{25}$; **12.** $\frac{3}{100}$; **13.** $\frac{45}{100} = \frac{9}{20}$; **14.** $\frac{2}{10} = \frac{1}{5}$. **B.** **1.** $\frac{7}{10}$; **2.** $\frac{85}{1000} = \frac{17}{200}$; **3.** $\frac{8}{10} = \frac{4}{5}$; **4.** $547\frac{15}{100} = 547\frac{3}{20}$; **5.** $\frac{3}{1000}$; **6.** $6\frac{625}{1000} = 6\frac{5}{8}$; **7.** $64\frac{1}{10}$; **8.** $\frac{25}{10000} = \frac{1}{400}$; **9.** $8\frac{5}{100} = 8\frac{1}{20}$; **10.** $\frac{1}{100}$; **11.** $45\frac{5}{10} = 45\frac{1}{2}$; **12.** $\frac{5}{1000} = \frac{1}{200}$

Page 74. **A.** **1.** $\frac{1}{3}$; **2.** $\frac{2}{3}$; **3.** $\frac{1}{9}$; **4.** $\frac{177}{700}$;

Page 75. **5.** $\frac{61}{300}$; **6.** $\frac{13}{150}$; **7.** $\frac{21}{400}$; **8.** $\frac{41}{800}$. **B.** **1.** $\frac{1}{6}$; **2.** $\frac{1}{24}$; **3.** $\frac{9}{40}$; **4.** $\frac{187}{300}$; **5.** $\frac{1}{16}$; **6.** $\frac{21}{250}$

Page 76. **1.** $27.79 (27.7875 rounded to the nearest hundredth); **2.** $9.62; **3.** $.98; **4.** 2.73 gallons ($2\frac{73}{100}$); **5.** $15.19; **6.** .434 foot ($\frac{217}{500}$); **7.** 1.35 inches ($1\frac{35}{100}$)

Page 77. **A.** **1.** numerator, denominator; **2.** numerator; **3.** 10; **4.** 100; **5.** 1,000; **6.** 10,000; **7.** reduced; **8.** decimal; **9.** fraction. **B.** **1.** .25; **2.** 4.4; **3.** $.16\frac{2}{3}$; **4.** $3\frac{3}{4}$; **5.** $\frac{1}{3}$; **6.** 1.25; **7.** $\frac{1}{16}$; **8.** $\frac{3}{20}$; **9.** .025; **10.** $14\frac{4}{25}$; **11.** 2.2.

Page 78. **C.** **1.** $\frac{3}{4}$; **2.** 18.2; **3.** $.33\frac{1}{3}$; **4.** $\frac{7}{8}$; **5.** $\frac{19}{150}$; **6.** $.16\frac{2}{3}$; **7.** $5\frac{1}{200}$; **8.** $\frac{17}{200}$; **9.** 1.75; **10.** .625; **11.** 3.6; **12.** $\frac{2}{3}$; **13.** 1.4; **14.** $9\frac{17}{20}$; **15.** .02; **16.** .5; **17.** $\frac{3}{200}$; **18.** $1\frac{3}{500}$; **19.** $2\frac{1}{3}$; **20.** 1.6; **21.** .04; **22.** $\frac{1}{100}$; **23.** 1.75; **24.** $.09\frac{1}{11}$; **25.** $9\frac{1}{4}$; **26.** $3.09\frac{1}{11}$; **27.** $\frac{11}{100}$

Page 79. **1.** $142, $26.65, $168.65; **2.** $150.40, $73.32, $223.72; **3.** $181.20, $181.20, $362.40; **4.** $304, $133, $437; **5.** $210, $47.25, $257.25; **6.** $168, $67.20, $235.20; **7.** $134, $25.15, $159.15; **8.** $203.60, $40.72, $244.32; **9.** $270, $151.95, $421.95; **10.** $194.80, $267.96, $462.76; **11.** $295.20, $44.28, $339.48

Page 80. **1.** $19\frac{2}{3}$%; **2.** 209.9% ($209\frac{9}{10}$%); **3.** $\frac{4}{5}$%; **4.** 12%; **5.** .19; **6.** 1.98; **7.** .0625; **8.** .037; **9.** .068; **10.** .6767; **11.** .004; **12.** 2.06125 (2.0613); **13.** 30%; **14.** 29.5%; **15.** 127%; **16.** $65\frac{1}{4}$%; **17.** 270%; **18.** .04; **19.** $33\frac{1}{3}$%; **20.** 2.5; **21.** .216

Page 81. **A.** **1.** 34%; **2.** $12\frac{1}{4}$%; **3.** 3.4% ($3\frac{4}{10}$%); **4.** $\frac{4}{5}$%; **5.** 116.3% ($116\frac{3}{10}$%). **B.** **1.** 13%; **2.** $6\frac{1}{5}$%; **3.** 1.8% ($1\frac{8}{10}$%); **4.** $\frac{2}{3}$%; **5.** 45.9% ($45\frac{9}{10}$%); **6.** $78\frac{1}{7}$%

Page 82. **A.** **1.** .23; **2.** .01; **3.** .092; **4.** 1.12; **5.** .103; **6.** .36; **7.** .08; **8.** 1; **9.** .0825; **10.** .09; **11.** 2; **12.** .3; **13.** .33; **14.** .155; **15.** 1.5; **16.** 2.25; **B.** **1.** .14; **2.** .03; **3.** .044; **4.** 1.47; **5.** .115; **6.** .18; **7.** .04; **8.** 2; **9.** .0575; **10.** .02; **11.** .124; **12.** .072; **13.** .08; **14.** .62; **15.** .47; **16.** .167; **17.** .9; **18.** .824; **19.** .002; **20.** 3.47

Page 83. **A.** **1.** .045; **2.** .0367 ($.0366\frac{2}{3}$, $.03\frac{2}{3}$); **3.** .5675; **4.** .088;

Page 84. **5.** .006; **6.** 1.125; **7.** .00375; **8.** .0525; **9.** .0733 ($.0733\frac{1}{3}$, $.07\frac{1}{3}$); **10.** .005. **B.** **1.** .0525; **2.** .05625; **3.** .184; **4.** .02125; **5.** .0075; **6.** 1.035; **7.** .00625; **8.** .0425; **9.** .0867 ($.0866\frac{2}{3}$, $.08\frac{2}{3}$)

Page 86. **A.** **1.** 75%; **2.** 10%; **3.** 7%; **4.** $16\frac{2}{3}$%; **5.** 40%; **6.** $6\frac{1}{4}$%; **7.** 32.5%; **8.** 136%; **9.** $68\frac{4}{5}$%; **10.** 19%; **11.** $48\frac{1}{3}$%; **12.** 386%; **13.** 240%; **14.** 70%; **15.** 9%; **16.** $233\frac{1}{3}$%; **17.** $466\frac{2}{3}$%; **18.** $76\frac{1}{4}$%; **19.** 21.4%; **20.** .4%; **21.** $22\frac{3}{4}$% ; **22.** 165%. **B.** **1.** 28%; **2.** 80%; **3.** 15% ; **4.** $33\frac{1}{3}$%; **5.** 90%; **6.** $2\frac{1}{5}$%; **7.** 68.5%; **8.** 115%; **9.** $25\frac{1}{2}$%; **10.** 43%; **11.** 225%; **12.** 71.4%; **13.** 6.3%; **14.** $903\frac{1}{2}$%; **15.** 47%; **16.** 33.3%; **17.** 50%; **18.** $5\frac{1}{4}$%; **19.** 85%

Page 87. **A.** **1.** percent; **2.** two, left; **3.** two, right. **B.** **1.** .78; **2.** 87%; **3.** .125; **4.** .08; **5.** 250%; **6.** $16\frac{2}{3}$% (16.67%); **7.** 6.25; **8.** .0075; **9.** 87.5%; **10.** 30%; **11.** .24; **12.** 170%; **13.** 2,078%; **14.** 1.12; **15.** 85.5%; **16.** 162.3%; **17.** 2.06

Page 88. **C.** **1.** .25; **2.** 70%; **3.** 1; **4.** 92%; **5.** .03; **6.** $66\frac{2}{3}$% (66.67%); **7.** .008; **8.** 37.5%; **9.** .0825; **10.** $.33\frac{1}{3}$ (.3333, $.3333\frac{1}{3}$); **11.** 22.5%; **12.** $81\frac{1}{4}$%; **13.** 2; **14.** 900%; **15.** .0075; **16.** 225%; **17.** .3; **18.** 33%; **19.** $.80\frac{1}{2}$; **20.** .002; **21.** 627%; **22.** 7.5%; **23.** 2.5; **24.** 470%; **25.** .0025; **26.** 25%; **27.** 40%

Page 90. The credit union offers the best loan arrangement.

Page 91. 1. 70%; **2.** 1,550%; **3.** 6%; **4.** $1,711\frac{1}{9}$%; **5.** $\frac{9}{50}$; **6.** $\frac{18}{25}$; **7.** $1\frac{3}{4}$; **8.** $2\frac{1}{10}$; **9.** $\frac{3}{1000}$; **10.** $\frac{27}{400}$; **11.** $\frac{41}{400}$; **12.** $\frac{9}{500}$; **13.** 14%; **14.** 1.4; **15.** $\frac{1}{20}$%; **16.** $\frac{47}{1000}$; **17.** b; **18.** a

Page 92. 1. 80%; **2.** 20%; **3.** 90%; **4.** 70%; **5.** 62.5%; **6.** 87.5%; **7.** 15%; **8.** 68%; **9.** $733\frac{1}{3}$%; **10.** 14%; **11.** $8\frac{1}{3}$%; **12.** $11\frac{1}{9}$%; **13.** 150%; **14.** 375%

Page 93. A. 1. $\frac{1}{2}$; **2.** $\frac{7}{25}$; **3.** $\frac{3}{50}$; **4.** $\frac{3}{100}$; **5.** $2\frac{1}{2}$; **6.** $\frac{1}{10}$; **7.** $\frac{1}{100}$; **8.** $1\frac{3}{4}$; **9.** $\frac{1}{4}$; **10.** $\frac{13}{100}$; **11.** $3\frac{1}{4}$; **12.** $\frac{9}{100}$. **B. 1.** $\frac{3}{5}$; **2.** $\frac{18}{25}$; **3.** $\frac{1}{50}$; **4.** $\frac{7}{100}$; **5.** $2\frac{1}{10}$; **6.** $\frac{9}{50}$; **7.** $\frac{1}{25}$; **8.** $1\frac{1}{4}$; **9.** $\frac{3}{4}$; **10.** $\frac{19}{100}$

Page 94. A. 1. $\frac{1}{200}$; **2.** $\frac{1}{125}$; **3.** $\frac{7}{1000}$; **4.** $\frac{9}{1000}$; **5** $\frac{37}{400}$; **6.** $\frac{7}{80}$. **B. 1.** $\frac{3}{200}$; **2.** $\frac{13}{400}$; **3.** $\frac{1}{250}$; **4.** $\frac{3}{500}$

Page 95. 1. .9; **2.** 60%; **3.** $\frac{7}{50}$; **4.** 12.5%; **5.** .0625; **6.** 2%; **7.** $\frac{1}{20}$; **8.** .35%; **9.** .005; **10.** 120%; **11.** .04; **12.** $\frac{1}{50}$; **13.** 26%; **14.** .1; **15.** $\frac{47}{100}$; **16.** $\frac{1}{25}$; **17.** $\frac{1}{250}$; **18.** 6%; **19.** .27; **20.** $\frac{1}{20}$%; **21.** $\frac{33}{50}$; **22.** .0075; **23.** $\frac{18}{25}$; **24.** .0025

Page 96. 1. Store A; **2.** B & B Furniture Mart; **3.** Store Two

Page 97. 1. b; **2.** c; **3.** b; **4.** a; **5.** 33; **6.** 84; **7.** 3.45; **8.** .14; **9.** d; **10.** a; **11.** b

Page 98. A. 1. a; **2.** b; **3.** b; **4.** a;

Page 99. 5. b; **6.** c; **7.** b; **8.** c; **9.** c; **10.** b; **11.** b; **12.** c; **13.** b; **14.** c. **B. 1.** c; **2.** b; **3.** b; **4.** b; **5.** a; **6.** b

Page 100. A. 1. 176; **2.** 17; **3.** $2.16; **4.** $156; **5.** $16;

Page 101. 6. 30; **7.** 2; **8.** 14.9985 (15); **9.** .09; **10.** 5.95; **11.** $49.50; **12.** $4; **13.** 1.89; **14.** 84; **15.** 65.6; **16.** .225; **17.** 1.25. **B. 1.** 12; **2.** $306; **3.** $41.25; **4.** $52.28 ($52.275 rounded to nearest hundredth); **5.** 10.08; **6.** 90; **7.** 30; **8.** 10.0005 (10); **9.** .05; **10.** 2.25; **11.** $49.50; **12.** 1.06; **13.** .75; **14.** 1.2

Page 102. A. 1. $2,700; **2.** $60.90; **3.** 20 questions; **4.** 1.8 pounds.

Page 103. B. 1. $1,037.50; **2.** $4.53; **3.** $1,800; **4.** $46; **5.** $1,530; **6.** 36% (35.71%); **7.** 660,250

Page 104. A. 1. $54; **2.** $71.40; **3.** $273.42.

Page 105. B. 1. $243.75; **2.** $720; **3.** $589; **4.** $172.38 (172.375 rounded to nearest hundredth); **5.** $345.51

Page 106. 1. 200; **2.** 48; **3.** $6.30; **4.** $.24; **5.** $.08; **6.** a; **7.** c; **8.** b; **9.** a; **10.** d (17.83355 rounded to nearest hundredth)

Page 107. 1. 20; **2.** 336; **3.** $84.80; **4.** $200; **5.** $2.25; **6.** $1,640.63; **7.** $4.45; **8.** $5,596; **9.** $56.65

Page 109. 1. 24 months; **2.** 50%; **3.** $36.75; **4.** $82.50; **5.** $80.50; **6.** $67 (no warranty)

Page 110. 1. 60; **2.** 100; **3.** 108; **4.** 48; **5.** 40; **6.** 250; **7.** 32; **8.** 28; **9.** a; **10.** c; **11.** d; **12.** b; **13.** a

Page 111. A. 1. 160; **2.** 40;

Page 112. 3. 72; **4.** 6,480; **5.** 24; **6.** $15; **7.** $21; **8.** $4.50; **9.** $20.25; **10.** 900; **11.** $14; **12.** $2,010; **13.** 198; **14.** 1,400. **B. 1.** 80; **2.** 120; **3.** 20; **4.** 352; **5.** 17; **6.** 250; **7.** $6; **8.** $45; **9.** 27; **10.** 800; **11.** 400; **12.** $120; **13.** 100; **14.** 32

Page 113. A. 1. $88,000; **2.** $60. **B. 1.** $280; **2.** $2.30

Page 114. 1. 5; **2.** $1.60; **3.** $16; **4.** 137.5; **5.** 80; **6.** b; **7.** c; **8.** a; **9.** a; **10.** d; **11.** a; **12.** d

Page 116. 1. $17,700; **2.** $18,000; **3.** $12,536; **4.** $15,000, **5.** 200

Page 117. 1. 80; **2.** 75; **3.** 50; **4.** $\frac{1}{4}$%; **5.** d; **6.** b; **7.** a; **8.** d; **9.** b; **10.** c

Page 118. A. 1. 75%; **2.** 25%; **3.** 5%; **4.** $33\frac{1}{3}$%; **5.** 70%;

Page 119. 6. 12%; **7.** $41\frac{2}{3}$%; **8.** 16.5%; **9.** 12.5%; **10.** 1.6%. **B. 1.** 15%; **2.** 20%; **3.** 75%; **4.** 87.5%; **5.** 25%; **6.** 60%; **7.** $16\frac{2}{3}$%; **8.** 2%; **9.** 60%; **10.** .5%; **11.** $66\frac{2}{3}$%; **12.** 1%; **13.** $66\frac{2}{3}$%; **14.** 50%; **15.** 5%; **16.** 30.3%; **17.** 70%; **18.** 30%

Page 120. A. 1. 15%; **2.** 40%. **B. 1.** 72%; **2.** 32.5%

Page 121. A. 1. 25%; **2.** $33\frac{1}{3}$%; **3.** 20%. **B. 1.** 5%; **2.** 20%

Page 122.   **1.** 25%; **2.** 25%; **3.** 2%; **4.** 60%; **5.** 12.5%; **6.** b; **7.** c; **8.** d; **9.** c; **10.** a; **11.** a; **12.** b

Page 124.   **1.** $144; **2.** $84; **3.** $54

Page 125.   **1.** b; **2.** c; **3.** c; **4.** d; **5.** b; **6.** a; **7.** b; **8.** d; **9.** a; **10.** c; **11.** a; **12.** c; **13.** b; **14.** c; **15.** a; **16.** d; **17.** b;

Page 126.   **18.** c; **19.** c; **20.** b; **21.** a; **22.** d; **23.** b; **24.** a; **25.** c; **26.** d; **27.** b; **28.** b; **29.** c; **30.** b; **31.** a;

Page 127.   **32.** b; **33.** a; **34.** b; **35.** d; **36.** c; **37.** c; **38.** b; **39.** d; **40.** c; **41.** d; **42.** b; **43.** a; **44.** b; **45.** b; **46.** a; **47.** b; **48.** d; **49.** a; **50.** c; **51.** a; **52.** d; **53.** a; **54.** c; **55.** a; **56.** d; **57.** a;

Page 128.   **58.** b; **59.** c; **60.** d; **61.** a; **62.** b; **63.** b; **64.** d; **65.** d; **66.** b; **67.** a; **68.** a; **69.** d; **70.** c; **71.** c; **72.** d; **73.** b; **74.** a; **75.** c; **76.** b; **77.** c; **78.** d; **79.** c; **80.** a;

Page 129.   **81.** b; **82.** b; **83.** b; **84.** c; **85.** d; **86.** d; **87.** a; **88.** d; **89.** c; **90.** b; **91.** d; **92.** c; **93.** a; **94.** b; **95.** b; **96.** b; **97.** c; **98.** b; **99.** c; **100.** a; **101.** b;

Page 130.   **102.** a; **103.** c; **104.** c; **105.** d; **106.** a; **107.** b; **108.** d; **109.** c; **110.** d; **111.** d; **112.** a; **113.** a;

Page 131.   **114.** d; **115.** a; **116.** b; **117.** a; **118.** a; **119.** c; **120.** d; **121.** a; **122.** b; **123.** d;

Page 132.   **124.** c; **125.** d; **126.** b; **127.** b; **128.** d; **129.** a; **130.** d; **131.** a; **132.** b; **133.** c;

Page 133.   **134.** c; **135.** b; **136.** c; **137.** b **138.** d; **139.** a; **140.** b; **141.** d; **142.** d; **143.** d; **144.** a;

Page 134.   **145.** d; **146.** a